NUTRITION AND FEEDING STRATEGIES IN PROTOZOA

Nutrition and Feeding Strategies in Protozoa

Brenda Nisbet, BSc, PhD,
Department of Zoology,
University of Aberdeen,
Scotland

CROOM HELM
London & Canberra

Croom Helm Ltd, Provident House, Burrell Row,
Beckenham, Kent BR3 1AT
Croom Helm Australia, PO Box 391,
Manuka, ACT 2603, Australia
Croom Helm Australia Pty Ltd,
28 Kembla Street, Fyshwick,
ACT 2609, Australia

British Library Cataloguing in Publication Data

Nisbet, Brenda
Nutrition and feeding in protozoa.
1. Protozoa
I. Title
593.1'041'3 QL366

ISBN-13: 978-94-011-6557-0 e-ISBN-13: 978-94-011-6555-6
DOI: 10.1007/978-94-011-6555-6

Biddles Ltd, Guildford and King's Lynn

CONTENTS

1 INTRODUCTION

Modern biologists describe protozoa as microscopic eukaryotic organisms with a capacity for establishing themselves in almost every conceivable habitat provided it contains moisture in some form. In 1674 at the time when Antony von Leeuwenhoek was making his first observations of 'very small animalcules' in Berkelse Mere near his home town of Delft, this concept of the ubiquity of protozoa would have been difficult to comprehend. Leeuwenhoek's curiosity later led him to examine the body fluids, gut contents and excreta of different animals and to describe 'an inconceivably great company of living animalcules, and these of divers sorts and sizes'. Here were early descriptions of parasitic protozoa, species which later came to be recognized as *Opalina, Giardia, Trichomonas* and others. Following his pioneering work in the field of microscopic observation, knowledge of protozoa has accumulated at an accelerating pace. Some 30,000 living species have been identified, and an equal number of fossil species, from habitats which range from the ocean waters to the exuvial fluid of insects. The study of protozoan nutrition is a particularly interesting aspect of this expanding field of zoology. What kind of nourishment do protozoa need, how do they acquire it, and what influence do the answers to these two questions have on where protozoa live?

The need to determine what kind of food protozoa are utilizing in their environment is desirable in all ecological studies involving microorganisms of aquatic communities. The ways and means of acquiring this information are difficult in practice. Laboratory cultures show what food is acceptable, but this does not necessarily reflect the natural food choice nor the food choice in the habitat under investigation. In some of the larger ciliates and amoebae it is possible to identify the contents of the food vacuoles in freshly-collected samples, but as Fenchel (1968) noted, vacuole contents very rapidly become unrecognisable except for the siliceous frustules of diatoms which retain their integrity.

Feeding Mechanisms

The ways in which protozoa acquire their food, by hunting, by para-

Figure 1.1: The Cilia and Subpellicular Infraciliature of *Euplotes*. (a) Section through the pellicle and a ciliary membranelle shows rows of three cilia with their kinetosomes and linking microtubules. Electron micrograph by L. Tetley. X 20,000. (b) A diagrammatic reconstruction of the dorsal and ventral patterns of subpellicular microtubules.

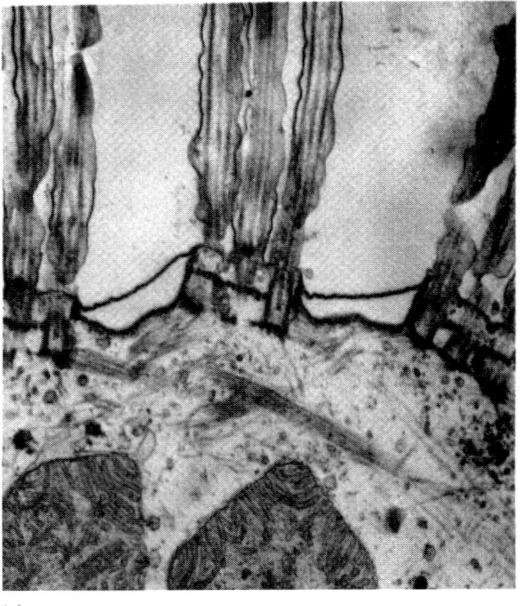

(a)

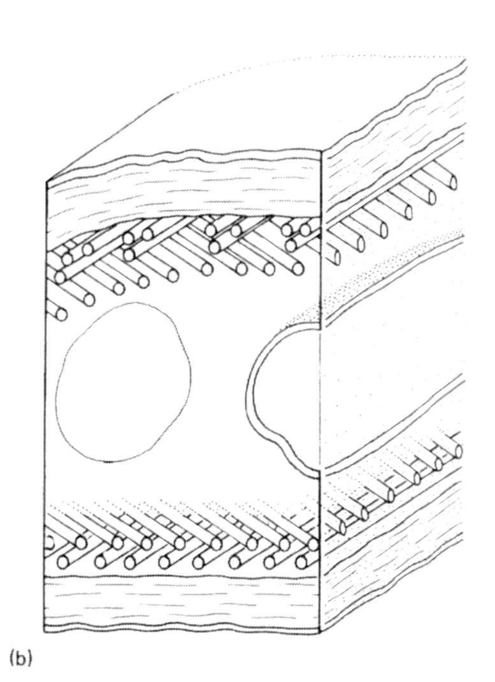

(b)

Source: Based on Grim (1967).

sitizing, by some form of scooping or vacuum-like suction, by chance contact or by photosynthesis will be discussed in later chapters. There is considerable uniformity in feeding behaviour within genera but this must not be assumed as a general rule. Feeding strategies depend partly on the potential for morphological adaptation which itself is related to the organization of the group. This point can be illustrated by considering a group such as the ciliates which show a wide range of morphological features based on the possession of cilia and the elaborate arrangement of microfilaments and microtubules which co-ordinate ciliary activity (Figure 1.1; and Corliss, 1979; Grell, 1973; Sleigh, 1973). The pattern of microfilaments linked to kinetosomes (ciliary basal bodies), which lies internal to the pellicle of a ciliate, makes up the infraciliature, now recognized as being of crucial importance in the study of protozoan relationships and evolution (Corliss, 1979). In spite of having an efficient, closely-structured infraciliature, the organism which depends on cilia for food trapping and movement is automatically restricted in both the size and type of food which can be taken.

The most generally accepted types of food – bacteria, algae and other small organisms – for free-living protozoa, are not specific to any one environment nor are they ingested by any one standard method. A ciliate uses ciliary currents to draw a stream of yeast cells towards its mouth; an amoeba engulfs yeast cells by pseudopodial flow and a flagellate may use its flagellum and a rod-like ingestion organelle. Different feeding strategies employed towards one type of food are illustrated in Figure 1.2.

The different food-trapping strategies just described are simply versions of small particle feeding in the broad scheme compiled by Yonge (1928, 1954). Yonge has three categories of feeding mechanisms based on particle size of food taken, which encompass all animals except the vertebrates. All three mechanisms, for trapping small particles, for engulfing large particles or whole organisms and for taking in fluids or soft tissues, may be found in protozoa (Table 1.1).

It will become evident, particularly from Chapters 8-12, that protozoan feeding mechanisms spread through these three categories irrespective of taxonomic relationships: 'Feeding mechanisms develop in correlation with the environment and the available food and any classification of them must cut clean across the subdivisions of the animal kingdom' (Yonge, 1928).

Even accepting this wide range of feeding mechanisms, it becomes clear that protozoa occupy a unique position in the biological world, because although the majority are structurally and nutritionally similar

Figure 1.2: Different Food-trapping Strategies towards One Type of Particulate Food (Yeast Cells). (a) *Paramecium* uses ciliary currents and an oral groove; (b) *Peranema* uses a probing flagellum and a rod-organ; (c) *Amoeba* ingests by pseudopodial flow.

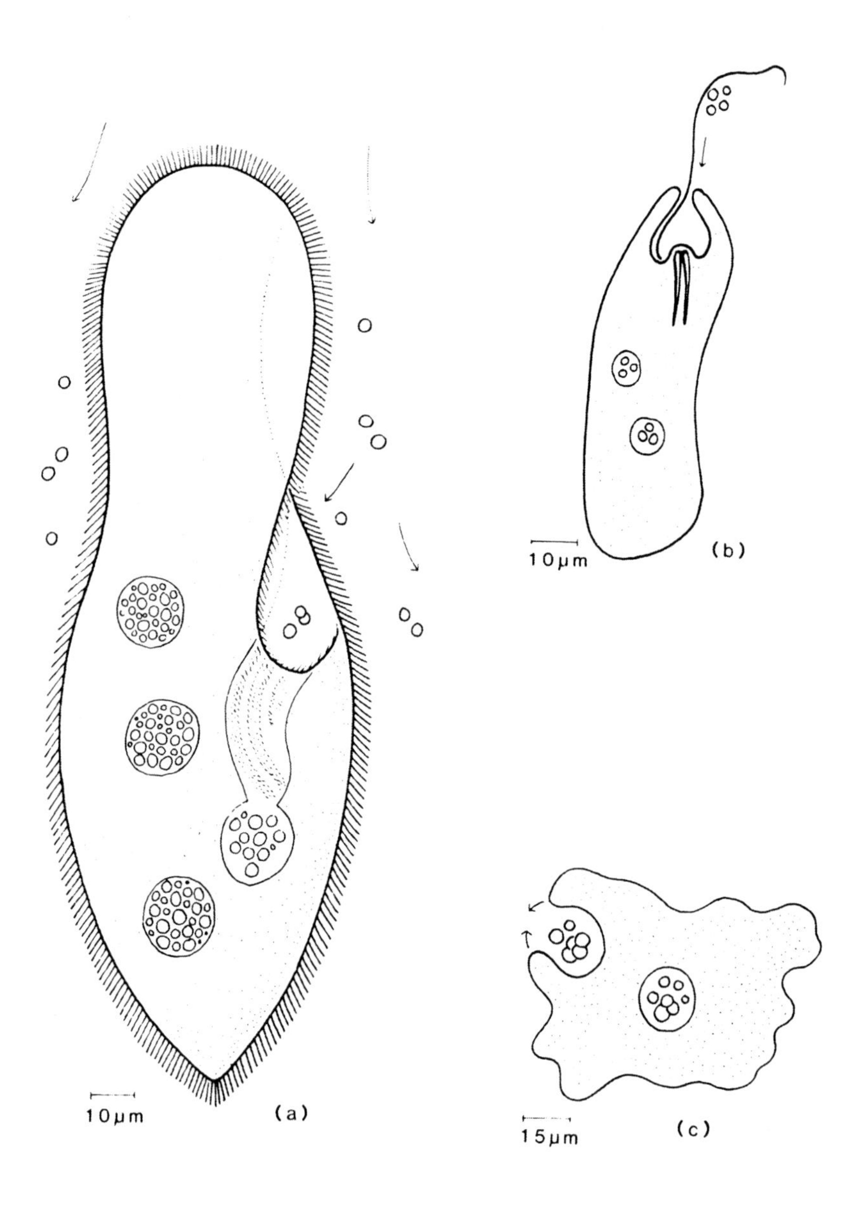

Table 1.1: Animal Feeding Mechanisms.

1. Mechanisms for trapping small particles involve the use of:
    - Pseudopodia
    - Flagella
    - Cilia
    - Mucus
2. Mechanisms for engulfing large particles or whole organisms:
    - Mechanisms for swallowing inactive food
    - Mechanisms for trapping the prey then,
        - swallowing whole
        - external digestion before swallowing
3. Mechanisms for taking in fluids or soft tissues:
    - Pinocytosis
    - Invagination of tissue fragments
    - Sucking

Note: Only those mechanisms which apply to protozoa have been included.

Source: Modified from Yonge (1928, 1954).

to animals, large sections of the Mastigophora (flagellated protozoa) include organisms with photosynthetic pigments and which have the nutritional physiology of plants. Given this characteristic then, it becomes necessary to justify their inclusion in what is essentially an animal phylum. Justification stems from the fact that many of the pigmented flagellates are closely paralleled by non-pigmented relatives which resemble them in gross morphology and movement, but which lack most of the cell organelles, chloroplasts, pyrenoids and eye spots, associated with photosynthesis. It is an inherent characteristic of protozoa that they are catholic in their feeding practices: the ability to adapt feeding to the type of food available is nowhere shown more clearly than in certain sections of the flagellated protozoa. *Euglena gracilis* is frequently quoted as a pigmented flagellate which, in the absence of sunlight as a source of energy, has to change from phototrophy to heterotrophy and rely on its ability to absorb suitable organic molecules such as acetates as a substitute for the products of photosynthesis (Chapter 4).

The boundary between microscopic plants and animals is difficult to define, especially within the mastigophoran flagellates. The conventional classification of flagellates which is being used here separates the photosynthetic flagellates and their close relatives (Phytomastigophorea) from the obligate heterotrophs (Zoomastigophorea), but accepts all as protozoa (see Chapter 9). Most modern textbooks of protozoology take this broad view (Farmer, 1980; Kudo, 1966; Sleigh, 1973). However, Hanson (1977), in his treatise on the origin and early evolution of animals, includes only some zoomastigophorans and excludes all phytomastigophorans, whilst other sections of the protozoa receive lengthy treatment.

Selectivity in Feeding

The majority of protozoa lie well within the boundary of animal-like in their feeding habits and their food choice is not too restrictive. Most predaceous ciliates will select prey on the basis of ease of capture (Güde, 1979). A potential food organism which is small enough or slow enough to trap easily is usually preferable to a larger vigorous prey. The amount of energy required to pursue this larger prey may not be balanced against the additional food content. An illustration of the wide range of foods acceptable to certain predaceous ciliates is shown in Table 1.2.

Table 1.2: Predaceous Ciliates and their Food Choice.

Ciliate Species	Food Choice
Chilophrya utahensis	bacteria, plant cells, amoebae, mastigamoebae, small ciliates
Coleps hirtus	ciliates, cryptomonads, green flagellates, bacteria
Dileptus gigas	ciliates, flagellates, amoebae, rotifers
Frontonia spp.	green algae, diatoms, desmids, *Euglena*, other protozoa
Sonderia vorax	diatoms, *Euglena*, rhodobacteria, *Oscillatoria*
Trachelius spp.	peritrich ciliates, algae, rotifers
Trachelocerca spp.	flagellates, ciliates, 'worms' and their eggs
Urostyla grandis	ciliates, flagellates, rotifers, diatoms

Source: Yonge (1954), Sandon (1932).

There is a small number of free-living protozoa which are extremely selective. These include the predator, *Didinium nasutum*, which feeds preferentially on *Paramecium* species. So close is this predator-prey relationship that *Didinium* depends on consuming forty-five well-fed paramecia a day to maintain maximum division rate (Butzel and Bolten,

1968). Several species of *Paramecium* are acceptable, but different species result in variation in cell volume of *Didinium*. When feeding on the larger *P. caudatum*, *Didinium* has an average volume of 9.1 x 10^5 μm^3; on the smaller species of *P. tetraurelia* the average volume is 5.6 x 10^5 μm^3 (Hewett, 1980). *Tetrahymena, Stylonychia, Blepharisma* and *Amoeba* have all been observed to show morphological variability according to food size.

Selectivity is also displayed by some suspension-feeding ciliates when they are presented with a range of acceptable food choices. In nature the ciliate would normally be faced with a mixed population of potential food organisms and probably any evidence of selectivity would be obscured by the density or differing 'catchability' of food. A food organism present in very large numbers is more likely to be consumed than one which is scarce although preferred. In an attempt to eliminate the effects of 'catchability' during feeding in the ciliate *Stentor*, Rapport, Berger and Reid (1972) designed a series of experiments in which *Stentor* was presented with various food organisms in single and mixed pair cultures. Two were described as protozoan food, *Tetrahymena pyriformis* and *Chilomonas paramecium*, and two as algal food, *Euglena gracilis* and *Chlamydomonas reinhardti*. Previously-determined standard densities of prey organisms were used (Rapport and Turner, 1970). As *Stentor* is a mainly sessile predatory ciliate which relies on motile prey swimming within the sphere of influence of its oral (food trapping) cilia, the density of the prey organisms is important. Given equal 'catchability' *Stentor* showed a definite preference for protozoan food, although continuing to take algal food in lesser proportions at the same time. Thus various levels of selectivity can be distinguished, but the majority of protozoa are capable of accepting a wide variety of food. Their feeding mechanisms, reproductive behaviour and general morphology show equal variability.

Unicellular Organisms

It is difficult to consider protozoa as unicellular organisms and yet how else can they be described? Acceptance of the term 'unicellular' in describing protozoa is perhaps an act of convenience. A unicellular organism, because of its nature, must be a more complex cell than one which forms a unit of a multicellular metazoan, and must be autonomous. Each protozoan has a full complement of enzymes for its metabolic and physiological life processes and, in addition, the appropriate sets of organelles for movement, food trapping, reproduction and osmoregulation all housed within its plasma membrane. Thus structural complexity, when compared to cells of multi-cellular organisms, is a most obvious and striking feature of protozoa whether at cellular or subcellular level.

Difficulty in categorizing protozoa as unicellular does arise with certain parasitic groups of myxosporeans, haplosporeans and possibly microsporeans, which may produce spores of apparently multicellular origin. This is one of the aspects developed in the recently revised classification of protozoa, the authors making a plea for the erection of seven separate phyla of protozoa (Levine *et al.*, 1980). Although protozoologists are aware of the miscellany of organisms assembled into one phylum they may not be ready to accept these major changes for common usage.

Classification

With the development of more sophisticated techniques and the increase in the number of protozoologists amassing further facts, it may be that the scheme of classification adopted here does not represent true relationships with the accuracy which its proponents are striving for. This they recognise: 'The present classification will, it is hoped, be used for many years by protozoologists and non-protozoologists alike, but it is inevitable that it will be modified as new information becomes available' (Levine *et al.*, 1980).

Classification of the larger groups is based principally on locomotory organelles, flagella, cilia, pseudopodia: their structure and their use or absence. Upon this framework then, further divisions are imposed according to the structure of the outer covering (pellicle, cellulose envelope, spore capsule) and the nutritional type. Detailed descriptions of groups and an explanation of the terms used in this classification will be found in the appropriate chapters. A simplified classification, which uses the new phylum structure based on the revision by Levine and his co-workers, will be adopted here.

Phylum I. SARCOMASTIGOPHORA. With flagella or pseudopodia (or both) for locomotion and often performing a double duty for food-trapping; single type of nucleus.

Subphylum 1. MASTIGOPHORA. One or more flagella for locomotion; nutrition autotrophic or heterotrophic.

Class 1. PHYTOMASTIGOPHOREA. Mainly autotrophic nutrition; with pigmented chloroplasts; heterotrophs without chloroplasts clearly related to pigmented forms.

Class 2. ZOOMASTIGOPHOREA. Chloroplasts absent; always heterotrophic; often parasitic.

Subphylum 2. OPALINATA. Oblique rows of short flagella over the entire body surface; nuclear division acentric; heterotrophic but cytostome absent; parasitic in the posterior digestive tract of amphibians.

Subphylum 3. SARCODINA. Pseudopodia for movement and food capture; ingest particulate food but at no specific point; naked or with protective shell or test.

Superclass 1. RHIZOPODA. Pseudopodia variable, lobopodia, filopodia, reticulopodia or flow with no distinct pseudopodium.

Superclass 2. ACTINOPODA. Mostly spherical and planktonic; axopodia supported by microtubules.

Phylum II. LABYRINTHOMORPHA. Form ectoplasmic, anastomosing networks; amoeboid cells glide over them; parasitic on marine plants and algae.

Class 1. LABYRINTHULEA. With the characters of the phylum.

Phylum III. APICOMPLEXA. Parasitic protozoa; sporozoites with an apical complex of rhoptries, conoid, polar ring; heterotrophs, some with a specific site for the uptake of subcellular particles.

Class 1. PERKINSEA. Flagellated 'zoospores' (sporozoites) with a conoid forming an incomplete ring.

Class 2. SPOROZOEA. Infective sporozoites with a complex conoid; reproduction asexual and sexual normally; mature trophozoites move by gliding or flexion with no external locomotor organelles except for some flagellated microgametes.

Phylum IV. MICROSPORA. Intracellular parasites; produce unicellular spores containing a uninucleate or binucleate sporoplasm and an extrusion apparatus with a polar tube and cap.

Class 1. RUDIMICROSPOREA. Spore with a simple extrusion apparatus; cap with polar tube ending in a funnel shape.

Class 2. MICROSPOREA. Spore with complex extrusion apparatus; long polar tube coiling round inside the spore wall.

Phylum V. ASCETOSPORA. Spore-producing parasites; spore is multicellular with one or more sporoplasms; no polar capsules or filaments; operculum present in some.

Class 1. HAPLOSPOREA. Spore with one or more sporoplasms;

spore wall may have tail-like projection.

Class 2. PARAMYXEA. Spore bicellular, one parietal cell and one sporoplasm.

Phylum VI. MYXOZOA. Spore-producing parasites; spores of multicellular origin; one or more polar capsules and sporoplasms; spore membrane with one to three valves.

Class 1. MYXOSPOREA. Spore with one or two sporoplasms; polar capsules containing coiled polar filaments for anchorage; usually two valves; parasitic in cold-blooded vertebrates.

Class 2. ACTINOSPOREA. Spores with several to many sporoplasms; three polar capsules with coiled polar filaments; membrane with three valves.

Phylum VII. CILIOPHORA. With simple or compound cilia at some stage in the life history; heterotrophs; well-developed cytostomes and food-trapping organelles in most groups; two types of nuclei.

Class 1. KINETOFRAGMINOPHOREA. Oral cilia only slightly distinct from body cilia; cytostome often apical or midventral, with or without a vestibule; cytopharyngeal apparatus often prominent.

Subclass (i). GYMNOSTOMATIA. Cytostome apical, subapical, or superficial; circumoral cilia simple; tubular cytopharyngeal apparatus (rhabdos type) strengthened by nematodesmata.

Subclass (ii). VESTIBULIFERIA. Vestibule usually apical or sub-apical, leading to cytostome; cytopharyngeal apparatus resembling the rhabdos type.

Subclass (iii). HYPOSTOMATIA. Cytostome on ventral surface; cytopharyngeal apparatus tubular and curved (cyrtos type), strengthened by nematodesmata linked into a 'basket' shape.

Subclass (iv). SUCTORIA. Adult sessile with suctorial tentacles; migratory larva ciliated.

Class 2. OLIGOHYMENOPHOREA. Oral apparatus and oral ciliature generally well-defined, consisting of a paroral membrane (hymenium) on the right side; membrane originates from a double row of kinetosomes in a zigzag pattern (stichodyad), with the inner kinetosomes barren and outer ones bearing cilia; on left a small number of compound ciliary organelles; cytostome usually antero-ventral in a buccal cavity.

Subclass (i). HYMENOSTOMATIA. Uniform body ciliation;

buccal cavity ventral or absent; kinetodesmata present; mostly freshwater forms, free-swimming.

Subclass (ii). PERITRICHIA. Somatic ciliature reduced to a temporary locomotor ring, mostly stalked, sedentary and often colonial; prominent oral cilia of paroral membrane and adoral membranelles.

Class 3. POLYHYMENOPHOREA. Conspicuous adoral zone membranelles often extending on to the body surface; somatic ciliature variable, may have cirri; often large; cysts common.

Subclass (i). SPIROTRICHIA. With the characters of the class.

Towards the Main Theme

After setting the scene on the protozoa in general, a method of grouping animals according to feeding mechanisms was considered. Its limitations are that it does not relate to the accepted taxonomic classification using morphological characters. It may be argued that in the context of this book categorising protozoa according to feeding would be more appropriate. This has not been done as the categories would be too large and too vaguely delimited, and would contain a vast assemblage of apparently unrelated organisms.

Protozoa must secure enough food, and of the right kind, to allow them to grow and reproduce. The division rate of the predatory ciliate *Dileptus anser* falls by 75 per cent when the concentration of its prey (*Tetrahymena*) is reduced ten times (Khlebovich, 1976). By studying the wide distribution and population sizes of protozoa it is self-evident that they have developed strategies for maximizing the use of the food available. In order to present these strategies here, two approaches are being developed: first, a review of the kinds of habitats in which protozoa thrive and the food available in them, and secondly, a description of the ways in which protozoa are adapted to acquire that food and process it. Autotrophy, heterotrophy and metabolic pathways as they relate to protozoa are described in brief to establish a general background to nutrition.

Protozoa are then taken phylum by phylum to show the range of feeding methods. In those protozoa which live as parasites, feeding and the food available vary from host to host, which necessitates some description of the life histories of parasites.

The final chapter looks at how extreme physical conditions affect protozoan populations and considers various aspects of inter-relationships in communities in general, drawing together ideas and themes which have emerged in the earlier chapters.

2 ENVIRONMENTS AND ECOSYSTEMS: FRESHWATER AND MARINE

Protozoa are found in a remarkable variety of environments. The range of habitats extends from the frozen soils of the polar regions to the hot spring run-offs from thermal areas, from the crystal-clear waters of oligotrophic alpine lakes to the richly polluted waters of sewage systems which operate on biological treatment. They are found in freshwater and in salt lakes, in any place which affords them the food and the physical and chemical conditions for which they are adapted. Perhaps one of the most bizarre situations is in the droplets of water trapped in the bracts of the flower heads of the wild banana plant, *Heliconia*, where the ciliate *Paramecium multimicronucleatum* can be found How it gets there is not clearly understood: *Paramecium* does not form cysts, and so the ciliate must be transported in the active state to the flower heads of the plant, perhaps by a browsing snail (Maguire and Belk, 1967).

Because of their small size, protozoa are able to live in association with a great diversity of plants and animals, developing either temporary or permanent relationships with their hosts. These can range from a casual relationship in which the protozoa are merely carried about by the host (for example, a colony of peritrich ciliates attached to the abdomen of a small freshwater crustacean), to an obligate association in which protozoa and host are necessarily dependent each on the other. Familiar examples of obligate associations are those of the cellulose-digesting flagellates and their termite and woodroach hosts, or the rumen ciliates found in the digestive tracts of ruminants. Full descriptions of associations and the levels of interdependence can be found in many standard works (Dogiel, 1965; Levine, 1972; Sleigh, 1973).

These differing habitats are possible only because of the variety of feeding mechanisms and the morphological adaptations related to them. Before considering these mechanisms, and the use protozoa make of the food they get, it is helpful first to review the kinds of habitats chosen to allow maximum exploitation of the resources available.

Every habitat contains a series of niches or microhabitats with varying characteristics. A single factor or several factors may determine whether a population of protozoa can establish in a particular niche for a limited period of time, and yet another set of factors may determine whether the population can be sustained over a longer period. Environmental variables which must be considered in relation to protozoan populations include food supply, which varies from location to location, from day to day and from season to season (Barker, 1942; Bick, 1973; Borror, 1968; Fenchel, 1968, 1969; Finlay, 1978, 1980; Finlay *et al.*, 1979; Goulder, 1971a, 1974; Small, 1973; Wang, 1928; Webb, 1956, 1961). A second important factor is the effect of competition for the same food (Brown, 1964; Gause, 1934; Gause *et al.*, 1934; Lee *et al.*, 1966a, b). Both fluctuating numbers of carnivorous predators and the availability of refuges where the prey is less likely to be detected, affect the chances of survival of the prey (Brown, 1940; Dewey and Kidder, 1940; Gause, 1934; Gause *et al.*, 1936; Luckinbill, 1973; Maly, 1978; Salt, 1967, 1968, 1974).

Finding Food

The aim of protozoa is to feed and grow, and having reached a critical size, to reproduce their own kind. This might be interpreted as implying a sense of purpose in protozoa seeking out those habitats in which the appropriate food is readily available. That *Paramecium aurelia* will congregate in a bacteria-rich area in preference to one of clearer medium can easily be shown by adding a drop of ciliates to a dish containing artificial pond water and a softened decomposing wheat grain. Within two or three hours most of the paramecia will be seen browsing on the bacterial growth covering the wheat grain. Later, the appearance of enlarged pre-dividing and dividing individuals would suggest that this aim has been achieved.

The initial exploratory movements made by paramecia newly introduced to a culture vessel give the impression that a food-rich microhabitat is encountered by chance. This may be so, but increasingly evidence suggests that many protozoa are actively attracted to their food. The ciliate *Woodruffia* always attacks *Paramecium* at its posterior end, which suggests that a chemical attractant is located there (Salt, 1967). Chemoreception is generally supposed to be a most powerful mechanism by which a predatory protozoon identifies possible prey (Seravin and Orlovskaja, 1977). By using chemical food models these workers demonstrated that chemoreception certainly operates in the ciliates, *Didinium nasutum, Dileptus anser, Lacrymaria olor, Coleps hirtus* and also in *Peranema trichophorum* and *Amoeba proteus*. There is no reason to suppose that a chemoreceptive response initiated by specific chemicals is restricted only to these examples.

Prey mobility may also be important in attracting the predator to its food. In an experimental system bacteria-feeding ciliates are seen to respond positively to oscillating particles simulating motile bacteria (Karpenko *et al.*, 1977). However, this positive taxis towards prey, promoted by movement in the water, does not normally result in a complete feeding response unless it is followed by the appropriate chemical stimulation.

Once browsing ciliates have found a favoured microhabitat, active swimming ceases and the ciliates are likely to remain there feeding until their rate of consumption of bacteria exceeds the rate of replacement by bacterial multiplication. Very dense populations of protozoa are normally essentially transient in character, for when the favoured food source is exhausted they either disappear or at least maintain a population density which is too low to be readily detected

The Transience of Communities

The transience of natural communities of protozoa has been recognized during many ecological studies over the years. The communities, or micro-biocoenoses, have as their basic ingredients protozoa, algae, bacteria and to a lesser extent, fungi. Picken (1937) identified two chalk stream communities in his study area, the blue-green association (characterised by *Oscillatoria*) and the sewage fungus association (characterised by *Sphaerotilus natans*). Algae and bacteria were equally important food sources for the ciliates present in both associations. Obvious fluctuations in the population occurred from time to time, particularly amongst herbivorous ciliates.

Fauré-Fremiet (1951a) followed the succession of ciliates colonizing a mat of the filamentous bacterium *Beggiatoa* and its associated free-swimming bacteria. First to appear was the bactivorous *Colpidium*, followed by the predators, *Lionotus* and *Tetrahymena vorax*, and finally the detritivores, *Paramecium* and *Coleps*, as the bacterial mat disintegrated (Figure 2.1).

The idea that integrated communities of protozoa develop in association with local concentrations of different types of food was further reinforced by Webb (1956). Saltmarsh pools with a bottom film of diatoms and bacteria support large populations of herbivorous ciliates such as *Chlamydodon mnemosyne, C. triquetrus, Chilodon calkinsi*

Figure 2.1: Temporal Relationships in a Ciliate Food Web. The starting point is a mat of the filamentous bacterium *Beggiatoa*. (1) Rapid growth of *Colpidium*, becomes prey for the predators *Lionotus* and *Tetrahymena vorax*, which then decline; (2) slower appearance of *Glaucoma* and *Chilodonella* which results in (3), disintegration in the mat of *Beggiatoa* to provide detrital food for *Paramecium* and *Coleps*.

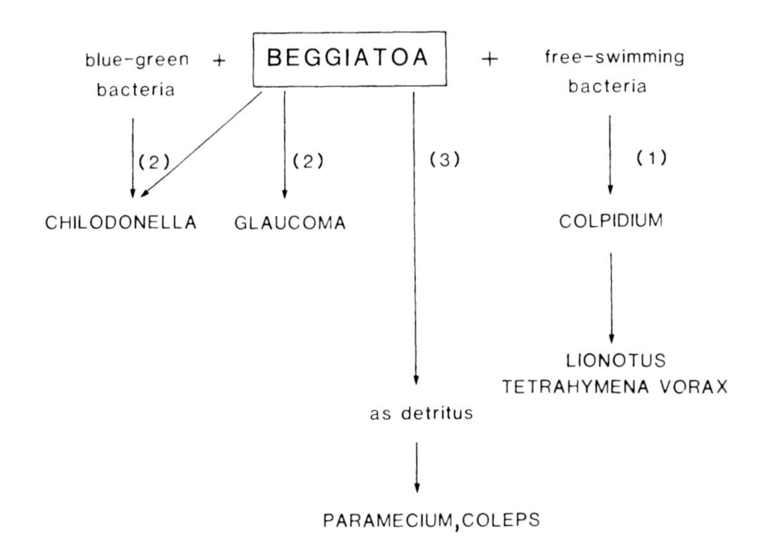

Source: Adapted from Fauré-Fremiet (1951a).

and *Frontonia marina*, particularly between May and September when algal growth is at a maximum (Figure 2.2).

Perhaps the most spectacular examples of transient communities are the blooms of autotrophic flagellates which flourish in conditions when a raised temperature coincides with a sudden influx of nutrients. Reviews of the occurrence of flagellate blooms, the species involved and their toxic effects may be found in Noland and Gojdics (1967) and in Taylor and Seliger (1979).

Changes in community structure, the factors which affect the rate of colonization and which species are the most successful colonizers in new habitats are problems interesting population ecologists in all fields of biology. Protozoa have come in for their share of attention. Henebry

Figure 2.2: Saltmarsh Ciliates and Their Food. (a) *Chlamydodon triquetrus*, (b) *Strombidium calkinsi*, (c) *Chilodonella calkinsi*, (d) *Uronychia transfuga*, (e) *Pleuronema marinum*, (f) *Cyclidium citrullus*.

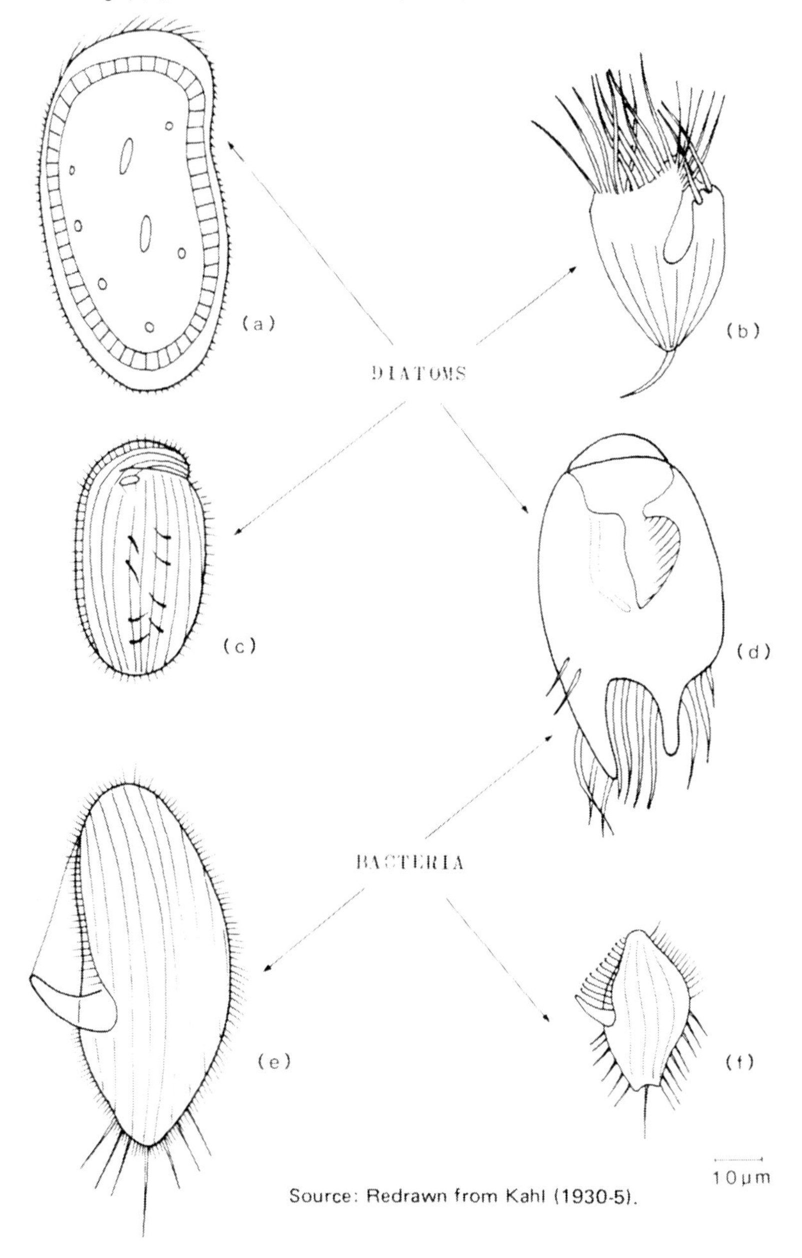

Source: Redrawn from Kahl (1930-5).

and Cairns (1980a) compiled a list of pioneer species of protozoa: species which most quickly colonized 'islands' in their laboratory micro-ecosystems (Table 2.1). Source pools, polyurethane foam units anchored in a pond, were allowed to accumulate protozoa for times ranging from three days to 13 weeks, before being transferred to laboratory systems. Three-day source pools, which would themselves contain a high proportion of pioneer species, contributed 100 per cent of their species to 'islands' in the test system. More mature source pools, up to 13 weeks old, contributed proportionately less of their species, and more slowly, as the population structure of the source pool was now more stable.

Table 2.1: Occurrence of Protozoan Species with High Pioneer Ratings on Source Pools Used in Microecosystem Experiments.

Species	Pioneer Rating	High Pioneer Value	Age of Source Pool 3 days	1 week	3 weeks	13 weeks
Monas sp.	4	X	X	X	X	X
Cyathomonas truncata	4	X	X	X	X	X
Chlamydomonas sp.	5	X	X	X		
Peranema inflexum	3	X	X		X	X
Ochromonas sp.	2	X	X			
Chilomonas paramecium	2	X	X			
Anisonema pusillum	2	X	X	X		
Cryptomonas erosa	3	X	X	X		X
Urotricha agilis	2	X	X	X	X	X
Cinetochilum margaritaceum	2	X	X	X		X
Glaucoma scintillans	2	X	X	X	X	
Cyclidium musicola	3	X	X	X	X	
Cyclidium litomesum	3	X	X			X
Hemiophrys sp.	1		X			
Mean No. of Species on Source Pools			14	31	36	55.5
No. Species with High Pioneer Value			13	9	6	7
Percentage of Species with High Pioneer Value			93	29	17	13

Source: Henebry and Cairns (1980a).

To be a successful pioneer species, a protozoon must have r-selection strategies, according to Henebry and Cairns (1980a). It must have a high reproductive rate, density-dependent mortality and the ability to use variable resources (Pianka, 1970). Protozoan colonization processes are discussed more fully in Cairns (1982).

Continuous flow systems have also been used to investigate community structure (Curds, 1966). Synthetic sewage passing through a laboratory activated sludge plant supports a changing population of protozoa as the sludge develops and the effluent improves in quality (Curds, 1966). Figure 2.3 shows the succession of dominant protozoa through time starting with colourless flagellates, followed by free-swimming ciliates, until the matured sludge after four weeks contains alternating peaks of creeping hypotrichs and attached peritrich ciliates. Using Henebry and Cairns' definition, the colourless flagellates are the pioneer species which are gradually replaced by a mature community.

Figure 2.3: The Succession of Dominant Protozoa in a Developing Activated Sludge. ●—●, flagellates; ■—■, free-swimming ciliates; ▲—▲, crawling hypotrichs; △—△, peritrichs.

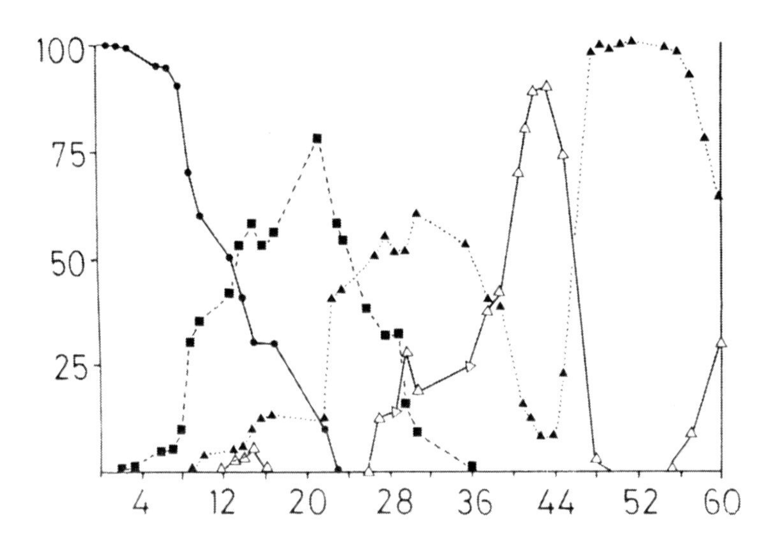

Source: Curds (1966), by permission.

Succession in protozoan species is related to changes in the types of potential food organisms available for capture. As the bacterial flora in an experimental sludge plant changes, the balance of protozoan species, which are predominantly bacteria feeders, will also change.

The ubiquity of protozoa in nature suggests that they must have a role, or at least an influence, however minor, in most ecosystems. The partnerships which develop between marine amoebae and bacteria in polluted ocean sediments result in extensive nutrient recycling and 'improvement' of the environment (Johannes, 1965). One can make estimates of the relative biomass of protozoa in particular ecosystems, but these may not be very meaningful in view of the considerable fluctuations in population which are encountered.

By considering the interactions in a variety of natural environments, particularly with respect to the micro-organisms present, it should be possible to formulate some idea of the complex inter-relationships which develop. The most obvious aquatic habitats, freshwater and marine, will be considered in this chapter, then the moist environments of soils and leaf litters and the more controlled decomposition and purification systems in the following chapter. This will result in some unavoidable overlap, but it may help to emphasize the more important aspects. Parasites, commensals and symbionts are given rather summary treatment, but will be considered more fully in later chapters.

The Freshwater Environment

Protozoa are an integral part of most freshwater ecosystems, although they make more impact on static bodies of water than they do on rivers and streams. This distinction between different types of freshwaters, the standing water (lentic) and the running water (lotic), is an important one to recognize. Running water presents problems for organisms which are too small to swim against the flow or have no particular adaptations for holding their position in the current. A more serious problem is the lack of food in such systems. The scouring effect of water flow reduces the likelihood of decaying organic material settling on the surfaces of stones and submerged vegetation and developing into a nutrient-rich film with bacteria and algae. Such a film normally forms the basis of a food web to support browsing protozoa: without it they will be absent. In contrast, settlement of detritus in still water can be extensive and even smothering at times. The difference between running water and the relatively static water of a lake is sufficient justification for considering these two aquatic macrohabitats separately.

The nutrient status of any body of water depends very much on its immediate environment of vegetation and soil. Minerals leach from the rocks and soil; increased nitrates and phosphates diffuse from agricultural lands where fertilizers are applied and cattle are reared. Excepting the localized discharge of industrial effluents, the major source of organic material is of plant origin, and so it is of prime importance whether the water is fringed by broad-leaved trees, conifers or herbaceous macrophytes. A stream running through arable land or a lake in open country will receive measurably less decaying leaf litter than similar bodies of water in deciduous forested areas. Comparison of the energy inputs from allochthonous (from outside the water) and autochthonous (from within the water) sources in streams flowing through open farmland and woodland with a heavily-overshadowing forest canopy produces a clear indication of the importance of marginal vegetation (Table 2.2).

Table 2.2: Comparison of Energy Sources in Freshwaters Expressed in $kJm^{-3}yr^{-1}$.

	River Thames (UK)	Bear Brook (USA)
Autochthonous		
algae and aquatic macrophytes	18,400	40
Allochthonous		
tree litter and macrophytes	515	11,210
groundwater, soluble and particulate organic matter	40,000	14,110

Source: Mann *et al.* (1972), River Thames; Fisher and Likens (1973), Bear Brook.

Energy budgets vary from stream to stream in response to local conditions and the data compiled for Table 2.2 illustrate an extreme situation in which the woodland stream (Bear Brook) may receive so much leaf litter that the small amount of algal growth adds little to the overall energy budget. In large eutrophic lakes airborne leaf litter contributes little to the organic carbon budget, although in small ponds or oligotrophic lakes its contribution is correspondingly more important (Gasith and Hasler, 1976).

Rivers and Streams

A river changes character markedly during its progression from the tumbling headwaters to the gently-flowing (but often polluted) lowland river. Carpenter's classification (1928) of British rivers into sections is still the most useful description: headstream, troutbeck, minnow reach and lowland reach. As these terms include both physical and biological characteristics it follows that not all rivers will necessarily include recognisable sections in all four categories. The headstream and the troutbeck sections of a river pose similar problems for protozoa. The rapidly-flowing variable water level and particularly the low nutrient status of a 'young' stream, which has not developed an appreciable bacterial and algal flora nor much organic detritus, create major barriers. In these conditions protozoan distribution is mainly restricted to isolated and perhaps temporary benthic microhabitats in quiet backwaters, if such exist, where detritus has accumulated.

The lowland sections of a river, the minnow reach and the lowland reach, present more sheltered conditions. The stony bottom of the minnow reach is now more stable and in many places is replaced by patches of silt where the water flow is reduced. Submerged macrophytes and mosses trap detritus and give shelter to grazing protozoa. The lowland reach, with its gentle flow and often luxuriant marginal vegetation, accumulates silt and detritus more rapidly and microbial activity increases. Many mesosaprobic protozoa, especially ciliates, can now colonise the benthos. Some organic pollution, with nitrates and phosphates, enhances algal growth and so increases the productivity of the habitat. Lower stretches of rivers often receive discharges of effluents from sewage works and industrial establishments. How these effluents affect the river depends on the amount of dissolved and suspended organic matter, the temperature of the discharge and the extent of dilution or the volume of effluent in relation to the size of the river. The way in which the character of the river may be altered for a considerable distance below the point of discharge will be discussed in the next chapter.

Lakes and Ponds

Lakes, ponds and even bogs are more hospitable environments for small organisms than are streams and rivers. Within a body of water four levels of habitat can be distinguished: the surface planktonic layer, the water column, the profundal zone and the littoral zone. By the profundal zone is meant those deeper parts of lakes, below the euphotic zone, where insufficient light prevents the growth of macrophytes. The littoral zone is essentially the shore line where submerged vegetation thrives along the margins or banks, a euphotic zone. If the body of water is shallow, light penetration and macrophyte growth will extend the littoral zone throughout the whole pond. Bottom deposits form the benthos of both zones. Profundal benthos readily becomes a community of anaerobic decomposers, subjected only to seasonal fluctuations, whereas littoral benthos is ever-changing, a community of aerobes which are continually mixed by water movements.

Without rigorous sampling techniques no clear idea of the spatial distribution of protozoa can be formulated. Dipping a jar into a pond has the merit of being simple, but will do little more than indicate whether the pond is 'rich' in animal and plant life, a useful starting point only. This abundance of animals may be subject to dramatic changes in oxygen and temperature at certain seasons of the year. This is particularly evident in temperate regions where during the summer months many eutrophic lakes develop thermoclines, temperature-stratified layers, consisting of an upper epilimnion which warms with the air temperature and is well-oxygenated and a lower hypolimnion which starts at about the lowest depth that light penetrates. With no photosynthesis and no water circulation, oxygen depletion in the hypolimnion is rapid and inevitable. In these conditions the ability of the population of protozoa to adapt to the changing environment becomes particularly important:

Plankton. Protozoa find their level where the appropriate food supply is available in quantity, and for most free-living protozoa this is not in the surface layers. This region is essentially the site for autotrophic flagellates, many quite spectacular species, others small and insignificant, but all having photosynthetic pigments ranging from yellow-brown to bright green. Common planktonic flagellates of freshwaters include the volvocids (*Volvox, Eudorina, Pandorina, Pleodorina* and *Gonium*), the dinoflagellates (*Ceratium* and *Peridinium*) and the chrysomonads (*Dinobryon* and *Synura*). Ciliates are not conspicuously present for they prefer to congregate at the site of maximum bacterial activity, which is in the benthos at the mud-water interface or amongst decaying detritus. However, there are a few heterotrophic plankton-dwellers amongst the sarcodines and these are adapted for remaining afloat with relatively little expenditure of energy. Heliozoans such as *Actinophrys* and *Actinosphaerium* employ two buoyancy devices to help themselves keep afloat: fine radiating pseudopodia (axopodia) and cell cytoplasm which is highly vacuolated. They prey on small flagellates which accidentally contact the axopodia. Others ingest filamentous or unicellular algae. Yet another group of minute amoeboid organisms, species of *Pseudospora*, invade cells of green planktonic algae and often occur in such large numbers that they may destroy more than 99 per cent of the algal population within two weeks (Canter and Lund, 1968).

Water Column. Protozoa living an independent existence are poorly represented in samples taken from mid-water in a lake. In that region the animal community (the nekton) is of larger and more mobile metazoans, which frequently act as transporters for epizoic peritrich ciliates or which house small heterotrophic flagellates. Although organic detritus and dead phytoplankton pass through the water column on their way to the benthic regions, this is not a site of great microbiological activity. In deeper lakes where the light intensity diminishes rapidly, autotrophic flagellates orientate upwards towards brighter light, leaving the nekton with a paucity of microbial life and protozoa in particular. Few choose to inhabit mid-water.

Profundal Zone. This zone of a lake normally supports a large and varied population of micro-organisms because here quantities of organic and inorganic nutrients accumulate. There are several sources of nutrients. Lakes may be thought of as sinks for the drainage water from surrounding agricultural land, from domestic and industrial developments and from open areas where leaching from rocks and soils may be considerable (Figure 2.4). Leaves and plant debris fall into the water, eventually settling in the benthos as fragmented, partially-decomposed organic sediments. The sequence of events leading to the decomposition of plant material is discussed more fully in the next chapter. In eutrophic bodies of water the dead remains of zoo- and phytoplankton add to the general nutrient pool after seasonal bursts of growth. In productive lakes of this kind where large quantities of organic material settle, anoxic conditions can develop below the thermocline in the summer months. If this happens the orange-coloured flocculant material of the oxidized microzone covering the mud surface rapidly changes to black anoxic mud (Gorham, 1958; Goulder, 1974).

A direct consequence of the onset of anoxic conditions is a change in the distribution of protozoa and other micro-organisms. Some protozoa will merely tolerate these conditions, whilst others, the true polysaprobes, will increase in numbers as they feed on anaerobic,

Figure 2.4: Nutrient Inputs to a Freshwater Lake.

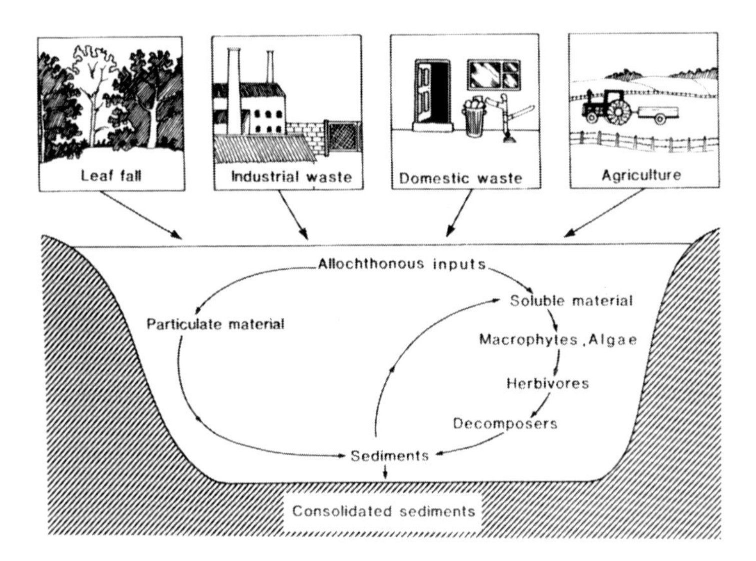

sulphur-reducing bacteria in the surface layer of the benthic ooze (Finlay, 1980; Goulder, 1971b; Webb, 1961). Amongst the genera characteristically found in anaerobic oozes where hydrogen sulphide may be produced in quantity, are *Caenomorpha*, *Metopus, Saprodinium, Epalxis* and *Pelodinium.*

The majority of ciliates, being mesosaprobes, cannot survive for long in conditions of very low oxygen tension in benthic sediments and respond by migrating upwards, following an acceptable oxygen level and food supply. This strategy can be expected to succeed in shallow lakes where benthic deposits become deoxygenated in summer but the overlying water circulates freely and remains adequately oxygenated. Consequently aerobic ciliates accumulate in the top centimetre of benthic sediment from May to September (Finlay, 1980). They increase in numbers here due to a sequence of events relating to higher water temperatures, to increased daylength boosting primary productivity, and to enhanced microbial decomposition in the sediment below. This latter activity uses up benthic oxygen and drives the redox discontinuity layer (the region of change from oxidizing to reducing conditions) nearer to the sediment surface. The relationship between the various factors which lead to a summer increase in ciliates at the sediment surface in shallow lakes is summarized in Figure 2.5, taken from Finlay (1980).

Figure 2.5: Increase in Number and Biomass of Ciliates in the Surface Sediment of a Shallow Lake in Summer.

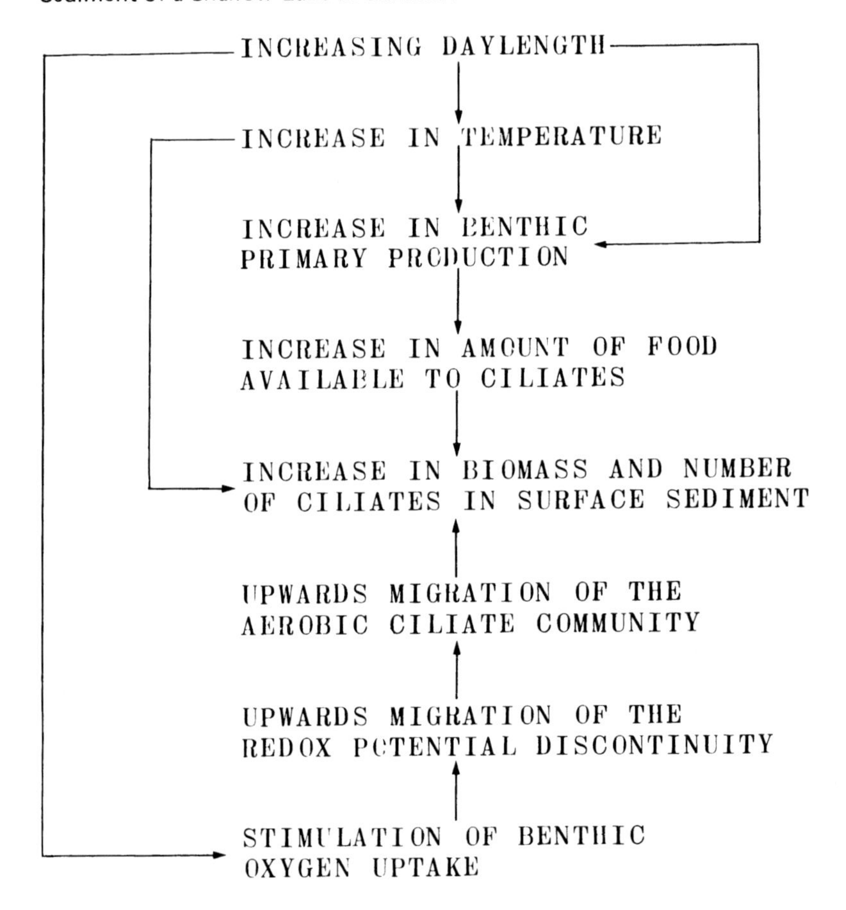

Source: Finlay (1980), by permission of Blackwell Scientific Publications Ltd.

Littoral Zone. Conditions in the littoral zone are often turbulent due to wave action and any plant detritus which falls in is not sufficiently decomposed to be useful as protozoan food. Thus the provision of a sheltered microhabitat becomes as important as in the upper stretches of a river. Submerged vegetation and the lower parts of emergent macrophytes offer useful sites for the attachment of many stalked and sessile protozoa such as the colonial peritrich, *Carchesium*, and some suctorians. *Stentor* and the suctorian, *Podophrya fixa*, have a preference for stones or fragments of detritus as settling sites. All have developed efficient mechanisms for trapping passing food, an essential attribute in non-motile organisms. Suctorians are opportunistic feeders, relying on their ability to retain a potential prey organism when it makes accidental contact with their tentacles (Figure 8.5). The peritrich ciliary apparatus is designed to draw a vortex of water towards the individual zooid, but if zooids are grouped into a branched colony of one hundred or more individuals then the combined effects of the ciliary currents will produce quite a substantial 'pull' (Figure 5.3).

In addition to providing shelter for littoral micro-organisms, marginal macrophytes and stones have the added bonus of collecting a film of bacteria, diatoms, filamentous cyanobacteria and detritus in all except extreme oligotrophic lakes where the nutrient status of the water is low. Hypotrich ciliates such as *Oxytricha fallax* with their 'walking' cirri, are particularly well-adapted for browsing on surfaces and probably take a variety of particulate food rather non-selectively. *Frontonia* and *Nassula* are grazers over the mixed growth, with *Nassula* selectively ingesting cyanobacteria filaments.

The foregoing pages have presented a brief review of freshwater habitats for protozoa, emphasizing the physical characteristics of the water in addition to nutrient status. In very general terms the most favourable conditions are to be found in the benthos of a shallow lake some one or two metres in depth, which is moderately eutrophic and shallow enough to escape thermal stratification and deoxygenation.

The Marine Environment

The marine environment is not as harsh as it would appear to be at first sight. It has particular problems as a habitat, but these appear to have been overcome by large numbers of protozoan species. High salinity in the open sea, variable salinity in estuaries, tidal rhythms, accumulations of mineral ions and decaying organic matter, especially seaweeds in the inter-tidal zone, are some of the factors which influence the environment. The inter-tidal zone is the most favoured, but the open sea also has its share of sites for micro-organisms to colonize. As with fresh waters, the general framework consists of littoral, planktonic and profundal (ocean floor) habitats, but within each habitat there exists a complexity not recognized in fresh waters (Borror, 1980; Sieburth, 1979). Sieburth discusses the distribution and relationships of micro-organisms in the sea and includes an extensive list of relevant publications. Only a brief introduction is necessary here.

Ciliates. Many studies of marine ciliate populations have been made, emphasizing particularly the relationships within ecosystems and the types of food which ciliates choose (Borror, 1968; Burkovsky, 1978; Dragesco, 1960; Fauré-Fremiet, 1950, 1951b; Fenchel, 1968, 1969; Hartwig, 1973, 1977; Hartwig and Parker, 1977; Smith, 1978). Although inter-tidal and brackish-water ciliates have attracted the most attention, for they are abundant and easy to sample, planktonic ciliates have their own special interest. Carnivorous and scavenging ciliates (gymnostomes such as species of *Prorodon* and *Didinium*) live in the coastal plankton where they find a ready source of food. The most common planktonic ciliates are the tintinnids which contain some 900 species in coastal and oceanic regions (Campbell, 1942; Fauré-Fremiet, 1924).

Sarcodines. The feeding behaviour and ecology of amoebae and their relatives in marine habitats have been described by Bovee (1973, 1979), Bovee and Sawyer (1979), Davis *et al.* (1978), Johannes (1965) and Sawyer (1971, 1980). All emphasise the importance of amoebae, particularly in marine sediment foodwebs, but recognize the difficulty of identifying small naked amoebae to the species level. Sawyer (1980) identified 26 species of amoebae from the marine sediments of the Western Atlantic Ocean and the Gulf of Mexico and tentatively identified several more. The testate or shelled amoebae are mainly confined to the benthos by their feeding habits, sluggish locomotion and lack of buoyancy (Lee, 1974; Murray, 1973). Those species which have adapted to a planktonic existence include members of the foraminiferan families, Globigerinidae and Globorotaliidae. *Globigerinoides ruber* capitalizes on light penetration in the surface layers of the sea by housing a population of symbiotic zooxanthellae, dinoflagellates similar to *Symbiodinium* species which contribute so importantly to the metabolism of coral polyps (Lee *et al.*, 1965). These workers consider that dependence on the photosynthetic activity of its symbionts keeps *Globigerinoides* in the euphotic zone, whereas symbiont-free *Globigerina bulloides* is not so constrained and is normally found deeper in the planktonic layers. The foraminiferans and other amoeboid organisms will be discussed more fully in Chapter 10.

Flagellates. Marine flagellates are many and varied, both morphologically and nutritionally. The dinoflagellates are responsible for much of the primary production by plankton in the euphotic zone. These distinctive flagellates are described in several early monographs and memoirs (Lebour, 1925; Graham, 1942; Kofoid and Skogsberg, 1928; Kofoid and Swezy, 1921; Tai and Skogsberg, 1934). Less conspicuous in the plankton, due to their smaller size, are other phytoflagellate groups – prasinomonads, prymnesiids, chrysomonads, cryptomonads and euglenoids. Much of the present knowledge of these small planktonic flagellates is due to Manton and her colleagues (Manton, 1964a, b; Manton and Parke, 1960, 1965; Manton *et al.*, 1965). Benthic flagellates may be pigmented and feed autotrophically (in shallow waters), but more commonly the benthos is populated by species which scavenge on bacteria and detritus in the bottom sediments. Large numbers of heterotrophic flagellates accumulate in areas of high decaying organic matter and high bacterial densities (Lackey, 1967).

Adaptation to the Changing Environment

The capacity of protozoa to adjust to changing salinities is a crucial factor in allowing colonization of the littoral, inter-tidal zone of the sea. Early experimenters demonstrated the extent of this physiological adjustment (Finley, 1930; Lackey, 1938, Yocom, 1934). Finley acclimatized 21 species of freshwater protozoa to sea water by gradually increasing the concentration. Some species, *Bodo uncinatus, Colpoda aspersa, Uronema marina*, *Pleuromonas jaculans* and *Cyclidium glaucoma* can be transferred directly from fresh water to sea water without detriment. The only adjustment which these animals have to make in these transfers is the slowing down or complete cessation of the contractile vacuole action, as this osmoregulatory organelle is no longer required when the external medium approaches isotonicity with their cytoplasmic fluid.

Estuarine protozoa are mainly euryhaline and adjust to decreasing salinity during ebbing tides by speeding up the action of the contractile vacuole. Fenchel (1969) considers that there are very few strictly stenohaline ciliates in marine littoral sediments and these few he has identified as belonging to the families Geleiidae and Loxodidae, which incidentally lack contractile vacuoles (Figure 2.6). These families contain marine psammophilic ciliates (excepting the freshwater *Loxodes*) and are members of the Karyorelictida, the less advanced ciliates (Corliss, 1979).

Marine Habitats

At the beginning of this section typical marine habitats were listed:

Figure 2.6: Marine Psammophilic Ciliates of the Families Loxodidae and Geleiidae, with Long Thread-like Bodies for Moving between Sand Grains. (a) and (b), *Remanella* spp.; (c), *Kentrophoros*; (d), *Geleia* sp.

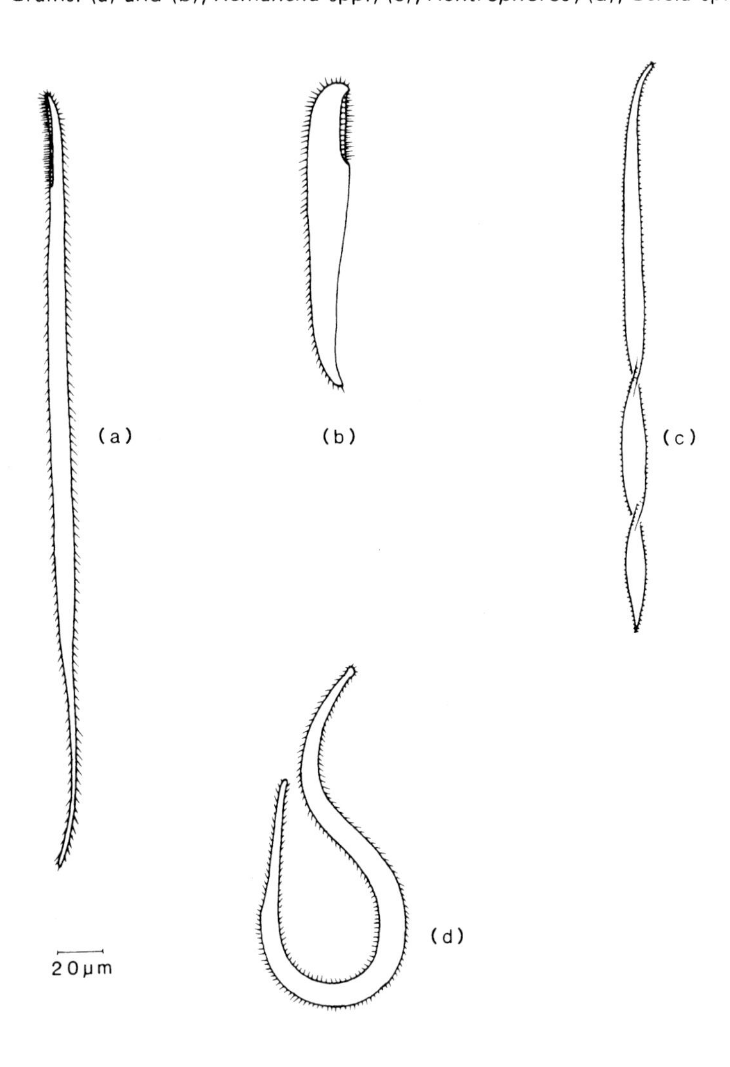

Source: Redrawn from Corliss (1979) and Fenchel (1969).

littoral, planktonic and profundal zones. The littoral zone is taken to include estuaries in addition to the shore line, which makes it more extensive and vastly more important as a habitat. By looking at the particular characteristics of each section of the marine environment one can build up a picture of what kinds of pressures operate against protozoa that live there and what kinds of food are available to them. What is there in one kind of habitat which allows large numbers and a wide diversity of protozoa to survive whilst another habitat has a very sparse and restricted population? Each major habitat will be considered with these questions in mind.

The Littoral Zone

All sea shores and estuaries are subject to fluctuations of temperature seasonally and diurnally and to tidal rhythms, yet this apparently unstable environment supports great diversity of animal life. However, micro-organisms which establish themselves in this changing environment can only build up sizable populations if other physical and chemical requirements are met. Many factors operate, with the most important being an adequate and renewable supply of nutrients, a suitable substrate and enough oxygen for aerobic organisms to carry out their normal metabolic activities. The possibility of desiccation is not a serious hazard as most protozoa can survive in the thinnest film of water for a considerable time, and can resort to limited downward migration following the water level until the tide floods in again.

Nutrients are carried down in the rivers, accumulate in inshore waters and are readily available for plant and animal growth. Incoming tides also renew the supply of nutrients, which may be inorganic salts for algal growth or the bodies of dead animals which supply organic compounds for bacterial growth. As the basic foods for free-living protozoa are normally bacteria or algae, then the areas of the shore where these micro-organisms can accumulate are likely to be favoured habitats. There have been several studies of littoral populations of protozoa in an attempt to relate population size with food supply. Fauré-Fremiet (1951b) observed that several species of the ciliate *Geleia* tended to congregate where purple sulphur bacteria were abundant in the top 10 mm of sand at Cape Cod. A major report by Fenchel (1969) describes localized communities which he calls 'sulphureta' where aggregations of sulphate-reducing bacteria, genus *Desulphovibrio*, grow anaerobically on decaying *Enteromorpha* and other seaweeds. This putrefying mass forms the basis of rather specialized microbial communities in sheltered bays and lagoons along the Baltic coast of Denmark.

Diatoms and bacteria are the main food of littoral interstitial ciliates, although some ciliates (*Geleia*) may prey on smaller ciliates (Fauré-Fremiet, 1950). This variety of food organisms, when it also includes dinoflagellates of the *Peridinium* type, will support a rich population of ciliates (Dragesco, 1960).

The importance of a suitable substrate cannot be overstressed. Several factors are important here in relation to the successful development of protozoan communities. It is clear from many investigations that the distribution of populations of interstitial ciliates is regulated as much by the sizes of the spaces between sand and silt grains as by the availability of food. Coarse sand has poor water-retaining qualities, drying out rapidly when the tide recedes, and allowing organic detritus to be washed out, whereas fine sands and silts remain water-logged with a thin film of water covering the surface, an advantage in allowing the continued growth of diatoms and other photosynthetic organisms. One problem in studying littoral habitats is that substrates may change over relatively short distances from the rocks of exposed shores to the fine sands and muddy silts of estuaries and sheltered bays where wave action is reduced. As might be expected, exposed rocky shores have few good habitats for protozoa, except in static pools around or above high water mark.

Sheltered habitats have been thoroughly investigated by Fenchel (1968, 1969), Fauré-Fremiet (1951b), Dragesco (1960) and others. Fauré-Fremiet, sampling beaches at Cape Cod, found that if the pore spaces between sand grains were less than 0.1 mm or were occluded with fine silt then ciliates were absent: pore spaces of 0.1-0.3 mm allowed colonisation by rather thread-like sluggish ciliates, often up to 2 mm long, the microporal ciliates. These ciliates belong mainly to the genera *Remanella*, *Geleia* and *Kentrophoros*, which also share other features in common (Figure 2.6). They possess strongly thigmotactic ciliated fields which allow them to form temporary attachments to sand grains as they glide through the substrate (Dragesco, 1960). Because of the restricted circulation of oxygen in these fine deposits microporal ciliates tend to be at or near the mud-water interface, never deep in the black anaerobic oozes lower down. Pore spaces of 0.4 mm allow a much more diverse population of mesoporal ciliates. Characteristic mesoporal ciliates are usually small and ovoid, displaying active jerky movements alternating with temporary thigmotactic attachments to sand grains (species of *Mesodinium*, *Placus*, *Remanella*, *Strombidium* and *Discocephalus*). A large number of species appear as occasional sand dwellers attracted by localized growths of diatoms or bacteria.

Foraminiferans become the dominant protozoan members of the meiofauna in some areas, their biomass exceeding that of the general meiofauna (Andren *et al*., 1968; Tietjen, 1971). Andren *et al*. sampled sublittoral sands of the coast of Göteborg, Sweden, and found that as they moved offshore the proportion of foraminiferan biomass to other meiofauna increased to 15:1 (Table 2.3).

Table 2.3: A Comparison of the Biomass of Foraminifera and Other Meiofauna in Sublittoral Sands off Göteborg, from Three Sites.

	Biomass (grams wet weight/m^2)		
	Inner site	Middle site	Outer site
Foraminifera	0.43	1.31	4.91
Other meiofauna	0.94	0.64	0.33

Source: Andren *et al*. (1968).

Using net weight as a measure of biomass for foraminiferans does not give an accurate idea of active organisms, as a large proportion of the organism is shell, and some shells are certainly empty. The highest biomass/m^2 quoted by Andren *et al*. (1968) was 4.91 g, as compared with Fenchel's (1969) figure for the biomass of ciliates on a Danish beach of 1.4 g/m^2.

These figures indicate clearly that protozoa are important constituents of the marine interstitial fauna. They consume bacteria and small algae and themselves become food for the invertebrate meiofauna. Reviews of marine interstitial communities by Gerlach (1971) and Swedmark (1964) both emphasize the importance of protozoa, ciliates and foraminifera in particular. The occurrence of small amoebae and bacteria-eating flagellates in nutrient-rich marine sediments has already been mentioned. Biomass estimations of these amoebae and flagellates are difficult: identification requires subculture of a sediment sample.

It has already been recognized that variations in oxygen tension in littoral deposits are related to the texture of the deposits, to the amount of oxidizable organic matter present and to the amount of nutrients available for algal growth. In many estuaries the receding tide leaves banks of fine sand and silt which are consolidated by blanket growths of diatoms and cyanobacteria. Diatoms migrate vertically within the top 2 cm of silt, causing a measurable variation in the amount of dissolved oxygen in the inter-tidal water. The surface water film may be

Figure 2.7: Micro-organisms in an Estuarine Food Web. Heavy lines indicate the principal sources of energy; light lines indicate pathways of utilization through micro-organisms; dashed lines show the sources of settled detritus.

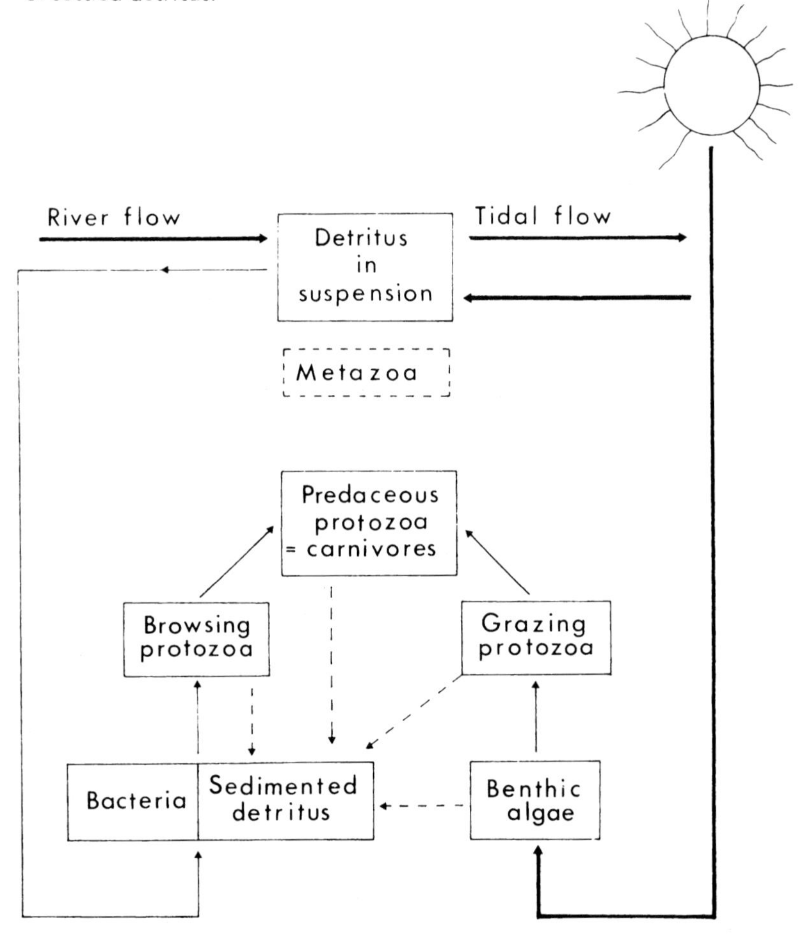

supersaturated due to the photosynthetic activities of diatoms (Aleem, 1949) with the oxygen content falling from 1.4 mg $0_2 l^{-1}$ at 2 cm to 0.3 mg $0_2 l^{-1}$ at 5 cm depth (Brafield, 1964). However, minor fluctuations in oxygen concentrations at the mud-water interface do not affect protozoa unduly, as most are relatively tolerant of slight reductions in oxygen tension. Beneath the surface layer anaerobic conditions prevail, and the silt is black with iron sulphide deposits resulting from the decomposing activities of sulphur bacteria. Only a restricted number of polysaprobic protozoa can tolerate these conditions and feed on such bacteria.

The importance of bacteria in food chains of estuarine and littoral sediments cannot be overstressed. The site of greatest microbial activity is the mud surface where heterotrophic bacteria use the available oxygen causing anaerobiosis in the underlying interstices. Measurements of bacterial populations in a typical estuary emphasise this concentration at the mud-water interface with counts of up to 4.6×10^8 per gram surface mud, 0.3×10^6 per gram in the overlying water and only 1.5×10^3 per gram in the inflowing seawater (Open University, 1978).

The complex inter-relationships which develop in the estuarine environment between different levels of micro-organisms and their energy sources have been summarized in Figure 2.7. An estuarine food web is not complete without the inclusion of metazoa: they remove nutrients by feeding at all levels and return nutrients to the detritus pool by their faeces and dead bodies.

The Planktonic Zone

The primary producers are the basis of any planktonic ecosystem. They make up a variable population of predominantly unicellular autotrophic organisms, cyanobacteria, green algae, diatoms, chrysomonads, coccolithophorids, cryptomonads and dinoflagellates. All are dependent on light and must remain within the limits of light penetration, which in the English Channel is reduced to 25 metres by excessive amounts of suspended solids in the water. In clear open oceans light may penetrate to 200 metres. Phosphates and nitrates, which are essential nutrients for planktonic growth, are carried down from land masses, making the inshore waters richer than mid-ocean. For this reason planktonic 'blooms', especially of dinoflagellates, tend to be an inshore phenomenon. However, bursts of growth of diatoms and dinoflagellates do occur in mid-ocean in the spring when light intensity and temperature increase, and may result in temporary depletion of nutrients. Their dead bodies sink to the benthos, often to remain there trapped by the thermocline, until vertical mixing recycles the nutrient-rich detritus in the autumn. This may result in an autumn bloom of phytoplankton. In deeper oceans nutrients may never be recycled except in well-defined areas of oceanic upwelling.

Phytoplankton, in terms of bulk, are the most important source of food for planktonic heterotrophic protozoa. Nevertheless bacteria are very widespread in the seas, although their distribution in the open ocean is normally restricted to accumulations around fragments of detritus, or on the surfaces of zooplankton and larger marine animals. As food for planktonic protozoa, however, bacteria are of limited use because of their scarcity.

All planktonic micro-organisms need aids to flotation if they are to remain in the photic zone without expending much energy on swimming against the density gradient. As the top few centimetres of the sea are the site of greatest activity during daylight hours, then most planktonic micro-organisms, whether producers or consumers, collect here. The normal protozoan heterotrophs of this planktonic zone belong to a few rather specialized groups, the tintinnid ciliates, foraminiferans, radiolarians, acantharians and the aberrant dinoflagellate *Noctiluca*, all of which are predominantly marine groups. Most are adapted for floating effortlessly in the photic zone by various devices (Figure 2.8). The tintinnids are pelagic ciliates of the open seas, excepting a few inshore species, such as *Tintinnopsis campanula* (Figure 2.8a), which is found along the coasts of Northern Europe. Tintinnids secrete protective, vase-shaped loricas, on which they cement sand grains and fragments of detritus. They look like minute aquatic helicopters, using the rotary motion of their adoral ciliary membranelles to provide uplift and, at the same time, to draw in food. *Tintinnopsis* feeds mainly on small dinoflagellates and silicoflagellates and so tends to congregate where these organisms are plentiful.

Only a few foraminiferans are planktonic (globigerinids and globorotaliids); most are slow-moving benthic organisms, crawling by means of pseudopodia and using their shells for protection. The most important planktonic foraminiferan, *Globigerina*, occurs in large numbers, particularly in warm-water plankton. *G. bulloides* is the only representative round British shores (Figure 2.8b). The organism floats readily with its bulbous calcareous shell armoured by fine calcareous needles, but when the animal dies the shell sinks to the ocean floor. *Globigerina* species are important constituents of the profundal ooze in which their calcareous shells are preserved as permanent records for dating geological strata. *G. bulloides* feeds heterotrophically, mostly on flagellates, which it traps in its fine network of pseudopodia, but the dominance of heterotrophy as a means of obtaining nourishment is not so evident in other planktonic foraminiferans. The existence of symbiotic zooxanthellae in species of *Globigerinoides* has been mentioned earlier in this chapter. In evaluating the relative importance of autotrophy (through the agency of symbiotic algae) and heterotrophy in foraminiferans, it has been argued that autotrophy is

Figure 2.8: Marine Plankton Adapted for Flotation. (a) *Tintinnopsis* – the rotary action of its adoral ciliary membranelle provides uplift; (b) *Globigerina*, fine calcareous needles aid flotation, fine pseudopodia trap food; (c) *Acanthometron*, oil-filled vacuoles in the central capsule and frothy extracapsular cytoplasm reduce its specific gravity; (d) *Ceratium*, long shell horns and oil droplets aid flotation, but the two flagella must keep vibrating; (e) *Noctiluca* adjusts the specific gravity of its cytoplasmic fluid for vertical movement.

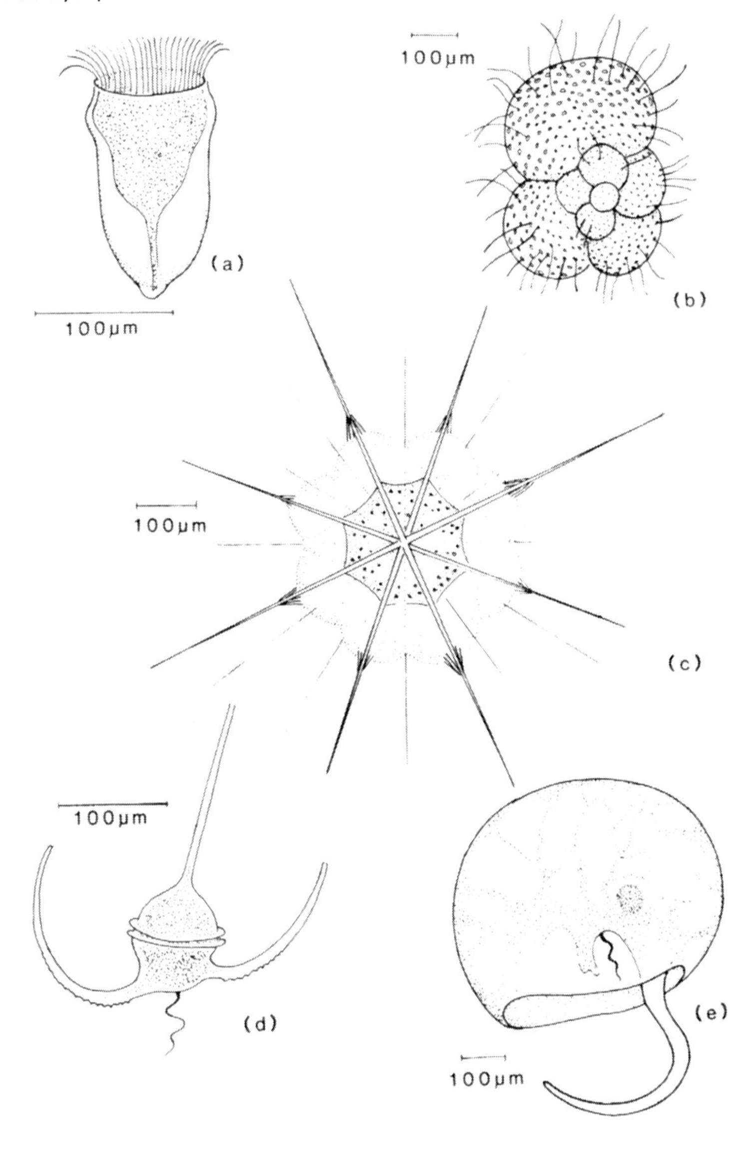

more important than heterotrophy (Lee *et al.*, 1965; Lee, 1974).

Radiolarians and acantharians are all planktonic, enclosed in their perforated capsules of silica or strontium sulphate secreted partly for protection and partly for support. They are important members of oceanic plankton, floating by means of oil-filled vacuoles in the intracapsular cytoplasm and watery vacuoles in the frothy gelatinous extracapsular layer. The typical acantharian pattern of twenty spines radiating from a strontium sulphate skeleton is shown in Figure 2.8c. Animals or plants of the plankton are trapped by long, pointed pseudopodia and drawn through the extracapsular calymma for digestion. Many species also house symbiotic zooxanthellae and can exist for long periods without particulate food provided that there is sufficient light for photosynthesis.

Dinoflagellates are essentially photoautotrophic and plant-like members of the plankton layer. One aberrant example, the carnivorous *Noctiluca miliaris*, floats near the surface of the sea where luminescent granules in its cytoplasm make a spectacular display in the evening light (Figure 2.8e). *Noctiluca* is an efficient 'fisher' using a sticky prehensile tentacle to transport larval crustaceans into its large cytostomal opening when it has made a fortunate catch, although it probably has to rely more on planktonic diatoms as a basic source of food. Buoyancy in *Noctiluca* is thought to be achieved by preferential accumulation of lighter ions in its cytoplasmic fluid rather than heavy divalent ones, although an alternative theory implicates the accumulation of hydrogen ions. Thus by some means the animal achieves a specific gravity less than that of sea water, but it can adjust this to facilitate effortless vertical migrations in the water column.

The Ocean Floor

The ocean floor, the profundal zone, as a habitat for protozoa, can be considered as an extension of the littoral benthic zone but with more extreme conditions. The total absence of light and, in many cases, a reduced circulation of sediments make for a very uniform environment. Although oxygen could become a limited factor in the very deep oceans, it rarely does as cold oxygenated water sinks from the polar regions. The supply of nutrients is steady. All the dead bodies of plants and animals in the oceans eventually form part of the profundal sediments. In spite of this sediment sink, the rate at which settled organic material is broken down by microbial action is very slow, as Janasch and Wirsen (1973) found. Packages of sterile nutrients left at 1,830 metres on the ocean floor for one year were decomposed at only one-third of the rate that similar material was decomposed in laboratory experiments maintained at an equivalent temperature (4°C) for one month. In shallower stratified seas there is vigorous heterotrophic bacterial decomposition which may cause oxygen depletion. Organic nutrients are released for use by other inhabitants of the benthos, or the bacteria themselves become a source of food. It is here that the majority of foraminiferans are found, species which are essentially deposit feeders, mopping up the remnants of algae and detritus, as they browse along the bottom, using their fine network of pseudopodia rather like a spider uses its web for trapping flies. Their shells remain for identification in ocean bed samples whilst other protozoa which may have been there escape notice. The rather specialised conditions in the deeper oceanic benthos make it a much less favourable habitat than elsewhere in the marine environment.

3 ENVIRONMENTS AND ECOSYSTEMS: SOILS, BIOLOGICAL PURIFICATION SYSTEMS AND OTHER ANIMALS

Many types of environment provide suitable habitats for protozoa: the list would be too long to discuss in any detail. The principal habitats of fresh water and the sea have already been considered in terms of physical, chemical and biological properties. Although both contain large and diverse populations of protozoa, other habitats can equal or even surpass them at times. Widening the scope to include the habitats of soils and plant litter and biological purification systems and to assess the types of relationship which protozoa establish with other animals will go some way towards covering the range of organisms in the protozoan phyla.

Soils and Plant Litter Habitats

Protozoa which choose to live in the soil have to contend with very variable conditions. They are subjected to the influences of certain abiotic factors: intermittent rainfall which affects the moisture content of the soil and the type of rock substratum which weathers in different ways. Factors dependent on the rock type are the pH of the soil water and the sizes of particles amongst which protozoa must find a suitable niche. Soil texture, or grain size, is clearly important enough to merit further consideration, as is moisture content. Of these factors only moisture content affects protozoa directly, the other factors operating more against alternative levels of the food chain.

Suitable sources of nutrients in the form of living or non-living organic matter determine whether a protozoan population can establish and increase in numbers. Most soils have at least a thin overlay of plant litter in varying states of decomposition, which is a steady source of organic matter passing into the food web. Thus the availability of plant litter detritus, coupled with bacteria, algae and fungi is significant. Decaying animal bodies will further add to the pool of organic nutrients.

Abiotic Factors and Soil Conditions

The amount of moisture in a soil has a profound effect on the type of community which develops in it. Therefore the texture of that soil becomes an important feature as texture determines water-retaining capacity. A coarse, sandy soil with large interstitial spaces has poor water retention and low organic content because much of the organic material is washed out of the surface layers. As the water drains downwards gases are able to permeate the pore spaces so there is ample oxygen for aerobic organisms, but without at least a film of moisture on the sand particles, micro-organisms will not survive in active form. At the opposite extreme a heavy, close-textured soil readily becomes waterlogged, oxygen cannot diffuse easily through the small pore spaces and rapidly becomes depleted. This situation restricts the activities of the aerobic decomposing micro-organisms on which many soil protozoa are dependent as a source of food.

Sources of Nutrients

Plant and animal detritus is of limited use to protozoa as a direct source of nutrient. It must be softened by bacterial and fungal attack and converted into bacterial protoplasm which may then be consumed by browsing protozoa.

Litter and Detritus. Leaf litter is recognized as being the major source of nutrient input to uncultivated soils, although agricultural soils receive additional organic and inorganic nutrients. Litter may be seasonal from deciduous trees shedding their leaves in autumn, leaves which have generally high decomposability, or from conifers or other evergreens, whose tough lignified leaves break down more slowly throughout the year. A carpet of pine needles will remain intact months after sycamore leaves have been reduced to skeletons. Partly through the cumulative activities of soil-dwelling bacteria and invertebrates and partly through the weather, decomposing organic matter finds its way into the soil. Evaluation of the part played by invertebrates (earthworms, nematodes, arthropods) as decomposers and as a source of animal detritus, is beyond the scope of this account, but it is important to realise that the surface layers of the soil are a site of great biological activity.

Bacteria. Because bacteria are the single most important source of food for the majority of protozoa, it is not surprising to find that those ciliates found most commonly in the soil feed preferentially on bacteria. The population of bacteria is closely bound up with the amount and type of leaf litter, for it can be seen that the more rapidly decomposable deciduous leaves and herbaceous litter support a larger population of bacteria than do the heavily-lignified and resinous conifer needles (Table 3.1).

Table 3.1: Bacterial Colonies Isolated from Deciduous and Evergreen Conifer Leaf Litter.

	Leaf Type	Bacterial Colonies (10^6/g dry litter)
Mulberry (*Morus rubra*)	deciduous	698
Red bud (*Cercis canadensis*)	deciduous	286
Pine (*Pinus echinata*)	evergreen	15

Source: Witkamp (1966).

However, the microbial community is essentially a cyclic one, changing not only with the type of leaf litter but also with the state of decomposition of the food resource, but few soil protozoa are selective enough to be greatly affected by a gradual change in bacterial species. The net effect of the decomposing activities of bacteria is to concentrate the proportion of nitrogen readily available to protozoa, for bacterial protoplasm has a C:N ratio of 10:1, as compared with the 25:1 ratio in deciduous leaves.

Algae and Fungi. Algae in the soil are less important as a direct source of nutrient than are bacteria. Their influence is more apparent in the establishment of a surface film over moist soil, a film which may become an important habitat with its characteristic microbial community. Poorly drained soils develop dense surface growths of algae, particularly cyanobacteria, and diatoms. Mat-like growths of cyanobacterial filaments, bound together by the mucilage they produce, are an important substratum for the grazing activities of protozoa, not so much as a source of food in themselves (for few protozoa feed preferentially on cyanobacteria) but as a site in which detritus and other micro-organisms accumulate. *Nostoc* forms large blue-green mucilaginous colonies easily visible to the naked eye, *Phormidium* and *Microcoleus* form blackish-green sheets on the surface of the soil (Figure 3.1). Mucilage production by diatoms is less copious, but nevertheless many small motile species such as *Navicula, Nitzschia* and *Pinnularia* can be a stabilizing influence on the surface of moist soils,

Figure 3.1: Common Algae and Cyanobacteria from the Surface of Damp Soils. (a) *Nostoc*; (b) *Microcoleus*; (c) *Phormidium*; (d) *Nitzschia*; (e) *Pinnularia*; (f) *Navicula*.

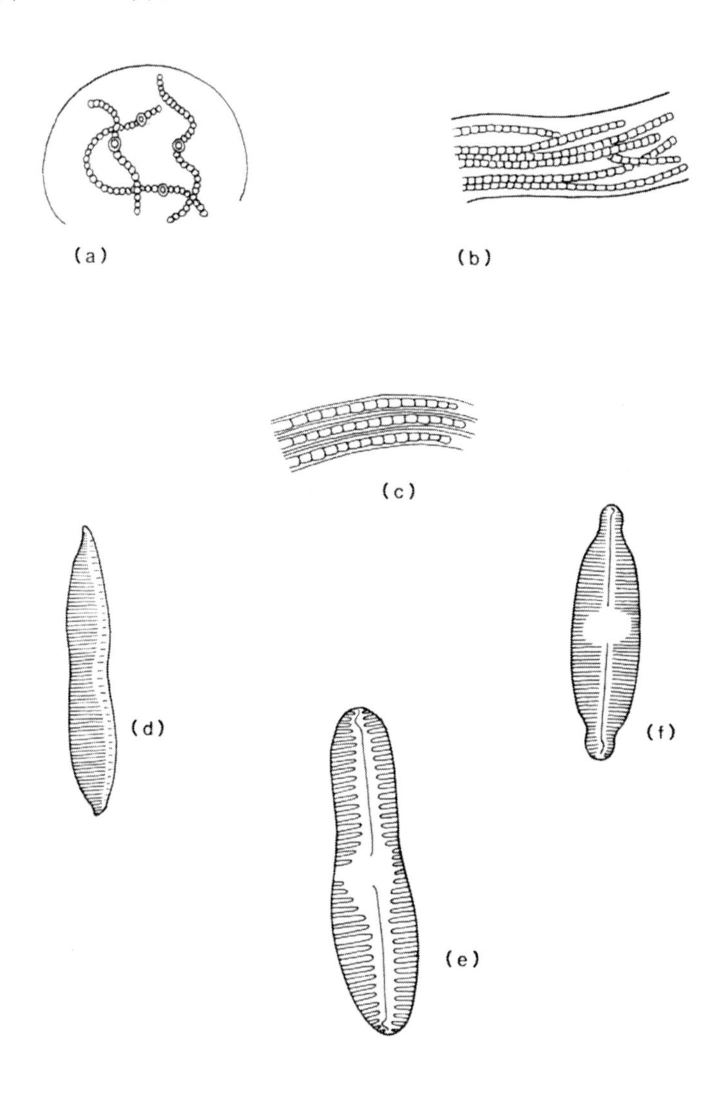

and are an acceptable, even preferred, food source for certain herbivorous protozoa.

Fungi and actinomycetes are not essential components of the protozoan diet directly, but they are important as active agents in the decomposition of plant and animal material. Any organisms which promote decomposition and encourage bacterial growth have a place in the soil food web.

Protozoa in the Soil

Of the 250 or so species of protozoa recorded by Sandon (1927) from the soil, very few – perhaps some 21 species – are exclusive to this habitat. The most commonly recorded species of flagellates, *Heteromita globosa*, *Oikomonas termo* and *Cercomonas* species, and ciliates, *Colpoda* species, are all small organisms capable of living in a thin film of moisture on the surface of soil particles and of feeding on bacteria. Several hypotrich ciliates, *Pleurotricha lanceolata*, *Gonostomum affine* and *Oxytricha* species in particular, are found in moist soils feeding on small algae and bacteria. They have the potential for successful soil dwellers, flattened flexible bodies for movement through a loosely particulate substrate, a readiness to form cysts (resistant for 3-4 years in *Oxytricha*) and a willingness to eat what is available, and yet they are not ubiquitous. Their distribution is limited by their larger size perhaps?

One of the most common and widespread ciliate genera found in soils and litters is *Colpoda*. Its capacity to respond to the moisture status of its environment by rapidly encysting or excysting, to multiply rapidly under favourable conditions, and its small size give it an advantage over many other ciliates (Stout, 1955). In view of this efficient adaptation it is surprising that Smith (1973c) was unable to find *Colpoda* in the soils of the maritime Antarctic islands, although it was present in sub-Antarctic South Georgia and in the Falkland Islands and in Arctic soils. Although the cold of the Antarctic winters would not affect *Colpoda* cysts, the cool short summers do not provide favourable growing conditions: *Colpoda* is unable to multiply fast enough to establish viable populations (Smith, 1973c).

Colpoda cucullus is ideally suited to edaphic habitats (of variable moisture status) provided the temperature is above the critical 4°C found by Smith (1973c). It has been shown to encyst, feed, grow and re-encyst within the space of eleven hours. In fact it can go through a full cycle of activity in the course of the night and encyst before the dew dries on the ground (Mueller and Mueller, 1970). And as repro-

Figure 3.2: Small Soil Ciliates Adapted for Scooping Bacteria into their Cytostomes. (a) *Colpoda cucullus*; (b) *Colpoda steini*; (c) *Colpoda maupasi*.

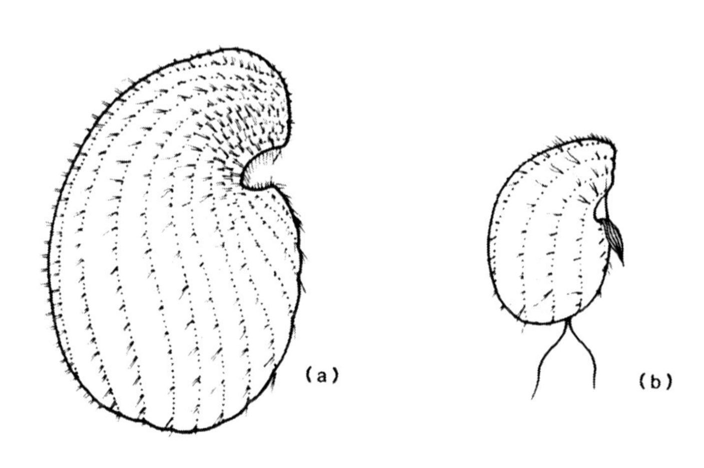

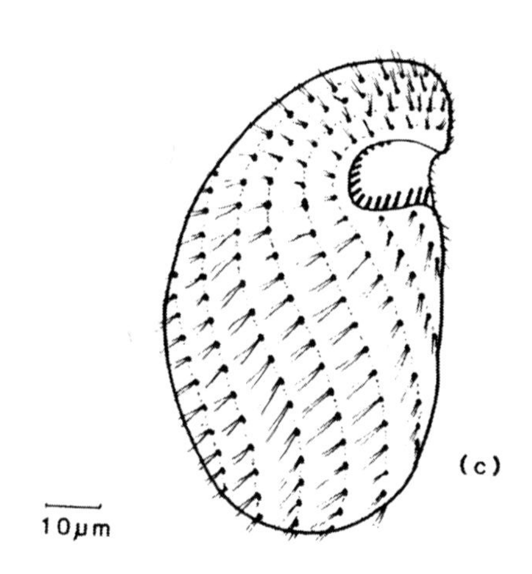

duction takes place within this cyst, at the next emergence up to eight individuals are released. In conditions of severe drought thick-walled resistant cysts are formed which are capable of surviving for five years or more. Three species of *Colpoda* are common in all soils and litters but nevertheless they do have particular habitat preferences. *C. cucullus* is more frequent in decomposing terrestrial vegetation, with *C. steini* and *C. maupasi* preferring moist soils (Bamforth, 1980). These small ciliates possess scoop-like cytostomes which operate as efficient collectors of bacteria during feeding (Figure 3.2).

Although the feeding effort of *Colpoda* spp. is directed towards collecting bacteria, *C. steini* will take cyanobacterium *Anacystis nidulans* if no other bacteria are available and will achieve a growth rate almost approaching that when fed on a diet of heterotrophic bacteria. Bader *et al*. (1976), using an experimental system, calculated that 1,000 *Anacystis* cells must be consumed to produce one new ciliate, which represents a feeding rate of one cell every 20 sec. This rate suggests that *C. steini* is able to collect unicellular cyanobacteria as efficiently as it does other bacteria, and utilize them.

Amoebae, particularly the testate amoebae, are major components of the permanent population of soil micro-organisms. The small naked amoebae are less obvious. Page (1976) describes twelve genera (23 species) of small naked amoebae commonly found in soils and leaf litter. Their cyst-forming habits and non-selective bacterial feeding are pointers to success. *Naegleria gruberi*, possibly the most common small amoeba, glides over the substratum 'by means of more or less eruptive, hyaline hemispherical bulges'. Its ability to transform temporarily into a biflagellate organism (non-feeding) in response to a sudden influx of water or to encyst for long periods in times of drought are obvious advantages. The many testate amoebae with their protective shells thrive particularly in mossy soils and acid soils, and leaf litter. They produce shells which are proteinaceous, siliceous, calcareous, or have foreign particles cemented to them. A shell may have a protective function, but it also restricts movement, except in soils where the pore spaces and the film of interstitial moisture are adequate. Large and spiny testaceans are generally restricted to bogs, very wet open-textured leaf litter or wet moss; small rounded testaceans may penetrate deeper into the soil (Figure 3.3). Small soil testaceans are rather opportunistic feeders, taking any bacteria, small algae, fungi and yeasts and selecting only for a size small enough to pass through the aperture of the shell.

Figure 3.3: Vertical Distribution of Testate Amoebae in a Moss and Soil Profile. (a) *Corythion dubium*, (b) *Euglypha* sp., (c) *Assulina muscorum*, (d) *Nebela* sp., (e) *Trinema enchelys*, (f) *Arcella discoides*, (g) *Euglypha strigosa*, (h) *Arcella vulgaris*, (i) *Nebela bigibbosa*, (j) *Difflugia globulosa*, (k) *Phryganella nidulus*, (l) *Trigonopyxis arcula*, (m) *Centropyxis hirsuta*, (n) *Cyclopyxis eurystoma*, (o) *Plagiopyxis* sp. Not to scale.

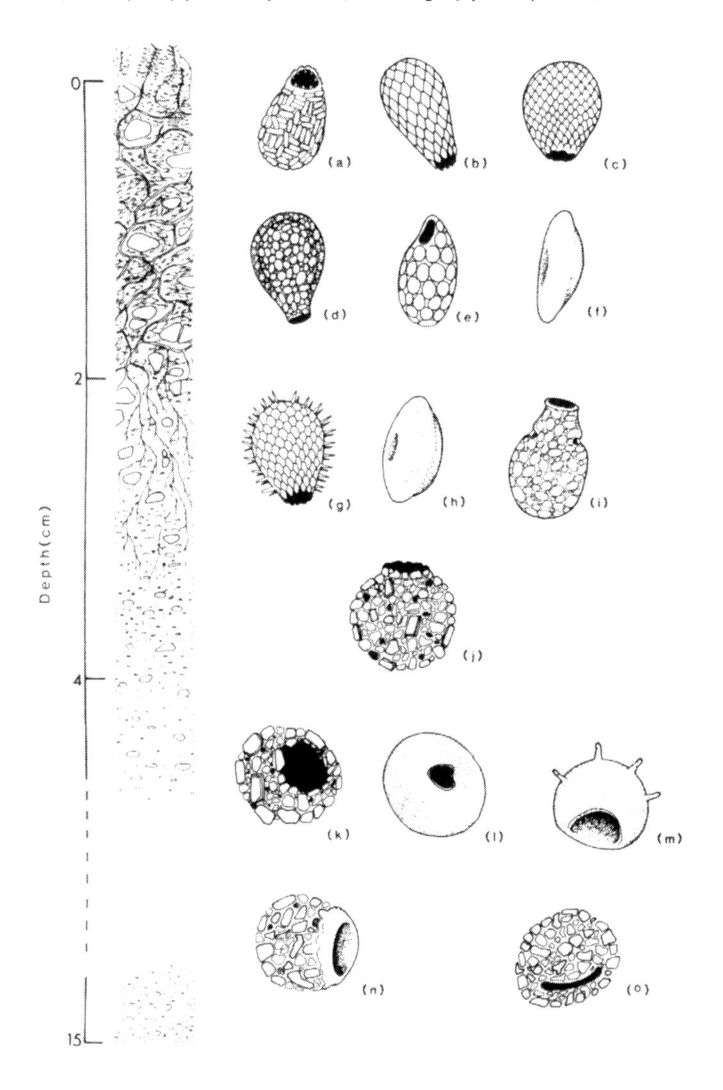

Source: C.G. Ogden, British Museum (Nat. Hist.), by permission.

Population Studies

Population studies on soil and litter ciliates and testate amoebae show clearly that there is an increase in populations of protozoa in a deciduous forest after autumn leaf fall, which follows the increase in bacterial numbers. Deciduous forest litter in winter supports 1,500 ciliates per gram, while the same site in summer has only 400 ciliates per gram (Table 3.2). Testate amoebae vary from 7,000 organisms per gram winter population to ten times fewer in summer litters. These figures also illustrate a point which has been noted by protozoologists interested in soil populations, that of the numerical dominance of testate amoebae in soils and litters. Motile ciliates are more dependent on a plentiful supply of suitable bacteria to provide energy for their activities; hence in areas of low microbial productivity such as coniferous forest floors and polar soils, testate amoebae predominate. These amoebae with their protective shells and slow-growing habits are favoured in areas of slow but continuous decomposition. Not surprisingly then, testate amoebae outnumber ciliates by about 23 to one in coniferous litter (Table 3.2).

Comparing the range of species of protozoa rather than population sizes in deciduous habitats, a different picture emerges. The figures from several deciduous habitats in the United States show that there is a bigger variety of ciliates than testate amoebae (although smaller total numbers), and that the surface layer of forest litter is a preferred habitat to the top few centimetres of soil (Table 3.2).

The soil environment is a complex pattern of microhabitats which are subject to large variations in moisture content and temperature. Add to these variables decomposing organic matter from a seasonal influx of deciduous leaf litter or a slow but continuing fall of evergreens throughout the year and the supply of nutrients and variety of microhabitats expands considerably. Protozoa are mainly restricted to the top few centimetres of the soil profile where they take advantage of the increased bacterial and fungal activity. Extreme fluctuations, particularly in moisture content, determine which protozoa can thrive in these habitats. They must be capable of moving in a thin film of interstitial water and when this dries up they must be able to encyst rapidly until external conditions promote a return of activity.

Decomposition and Biological Purification

Large amounts of decomposing organic matter are normally found in

Table 3.2: Number of Protozoa and Range of Species in Litters and Soils, Related to Seasons.

	Total[1] Species	Range of Species[1]		Protozoa (no./g)			
		Deciduous litter	Soil	Deciduous litter[2] Summer	Deciduous litter[2] Winter	Conifer litter[1]**	Soil[1]*
Ciliates	48	7-22	5-14	400	1,500	300	150
Testate amoebae	34	5-16	2-12	700	7,000	7,000	1,010

Notes:
* Data averaged from soil samples of ten deciduous forest sites; separate winter and summer figures not available.
** Average of nine conifer forest sites.

Source: (1) Bamforth (1971); (2) Stout (1962).

lakes and rivers and the breakdown of this material proceeds steadily, without unduly upsetting the natural population of micro-organisms. Only if organic matter is introduced in excess or contains toxic substances does a major change in population occur. That organic detritus, whether autochthonous or allochthonous, is an important source of energy can be seen from the data in Table 2.2.

Source of Nutrients

In bulk terms most of the allochthonous organic material will be leaf litter and other plant fragments, but as a rapid energy source it is inferior to the input of soluble organic compounds from farm drainage, sewage effluents, and even some trade effluents. These soluble compounds may be used directly in protozoan metabolism whereas utilization of plant detritus involves the activity of primary decomposers, bacteria and fungi, which 'soften' resistant cellulose and lignin skeletons. However, plant detritus is made available in a two-stage process on a time scale which relates to local conditions of temperature and pH, in that 15 per cent of the organic matter is leached out into the water quite rapidly, providing food for small flagellates and ciliates which rely on pinocytotic uptake or membrane transport mechanisms for their supply of nutrients. After this initial leaching, bacterial activities continue the breakdown process. Fenchel (1972), studying the decomposition of artificially-fragmented *Zostera* detritus, found that up to 25 per cent of the detrital surface was colonized by bacteria. Conceivably then, when protozoa are browsing over detritus they are engulfing large numbers of bacteria with no additional effort. This fortuitous energy-rich food source is of more value than plant detritus. Although readily utilizable, the high soluble and particulate organic content of sewage and trade effluents may not be advantageous to an aquatic system. It is important that a change of state in the ecosystem should be quickly and easily recognised. To this end various biological indicators of pollution have been formulated, but few are applicable to micro-organisms.

Assessment of Pollution

The amount of soluble organic matter and the oxygen needed to oxidize it are linked closely as important factors in assessing the degree of pollution of a body of water. Kolkwitz and Marsson's Saprobien System (1908), based on these two factors, remains widely favoured by freshwater biologists as a means of describing water and the organisms which live in it. The region of maximum pollution or the *polysaprobic* zone is characterised by high organic content and bacterial activity with very low dissolved oxygen, or completely anoxic conditions. The *mesosaprobic* zone includes a range of conditions which are more conveniently sub-divided into α-mesosaprobic, with moderate dissolved oxygen and bacterial activity, and β-mesosaprobic where dissolved oxygen is usually above 50 per cent saturation and bacterial activity is reduced due to less organic matter. The *oligosaprobic* zone then represents the unpolluted water, high in dissolved oxygen, as the low organic content and low decomposing activity make few demands on the ecosystem. Application of the Saprobien System is aided by identification keys which include saprobic category (Bick, 1972; Curds, 1969).

Changes of populations of protozoa and other micro-organisms as a result of pollution by effluent discharge into rivers and streams have also been categorized by Patrick (1949, 1950, 1961). She divided water courses into four habitat types; (a) healthy, with a wide variety of species; (b) semi-healthy, with certain species in disproportionately large numbers; (c) polluted, with sensitive species disappearing and the more tolerant species becoming more abundant; (d) very polluted, with extreme reduction in diversity, and a few species very abundant. All these habitats are the result of varying degrees of organic pollution. Patrick's scheme and the Saprobien System of Kolkwitz and Marsson as methods of classifying waters do not coincide precisely, but they do both express the same ideas – that increased organic pollution leads to an increase in population size but a decrease in diversity of those species which can tolerate a high level of organic matter, a reduction in oxygen tension and which feed preferentially on certain types of bacteria (Figure 3.4).

Protozoa in Decomposition and Purification

Protozoa cannot be considered in the dominant role in decomposition and purification systems, although any organisms present in an ecosystem in appreciable numbers do have some effect on that system, however minor. It is appropriate at this point to list briefly the various possible influences of protozoa that have been suggested. These can be grouped under the following general headings: mechanical agitation, stimulation of bacterial activity, consumption of bacteria and competition for nutrients.

Mechanical Agitation. This role is probably untenable as individual protozoa are too small to act as individual agitators which stir up detritus deposits. Only in the relatively massive colonies of peritrich ciliates

Figure 3.4: Assessment of Pollution Using Different Criteria.

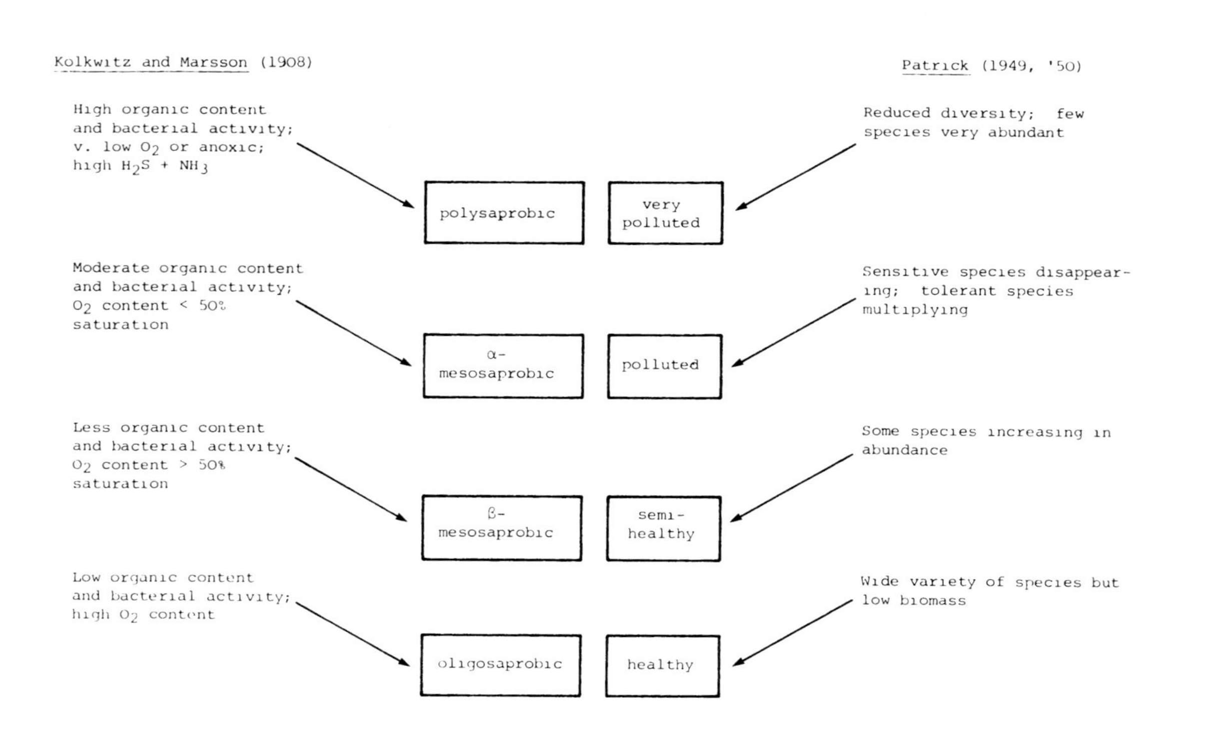

can any real mechanical agitation be demonstrated (Figure 3.9).

Stimulation of Bacterial Activity. Grazing protozoa may create micro-turbulence around growths of bacteria, reducing their density and hence reducing competition for dissolved nutrients and oxygen in the immediate vicinity of the remaining bacteria. This theory and the possibility that protozoa may release a growth-promoting substance have been proposed by Fenchel and Harrison (1976) in explaining why, in experimental systems, bacterial decomposition of hay detritus proceeds more rapidly in the presence of grazing protozoa (Harrison and Mann, 1975; and Figure 3.5). Grazing activity keeps bacteria in 'a prolonged state of physiological youth' (Johannes, 1965).

Figure 3.5: Rate of Bacterial Decomposition of Hay in Sea Water in the Presence and Absence of Protozoa: ▲—▲, bacteria only; ■—■, with choanoflagellates added; ●—●, with mixed protozoa.

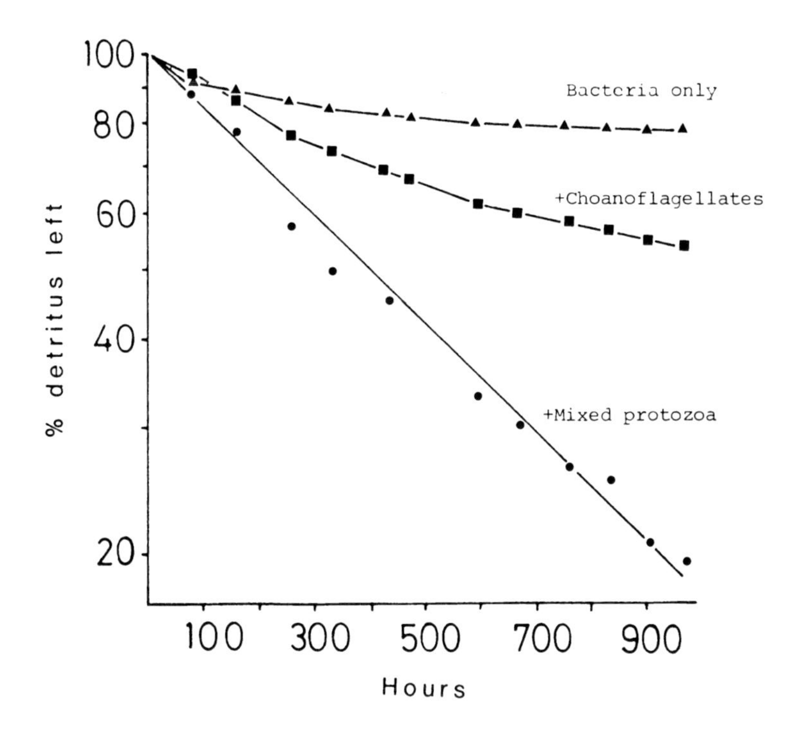

Source: Fenchel and Harrison (1976), by permission of Blackwell Scientific Publications Ltd.

Consumption of Bacteria. Protozoan grazing is a significant factor in controlling bacterial numbers in decomposer and purification systems. Evidence to support this role for protozoa is overwhelming: only a few selected reports are quoted (Barsdate *et al*., 1974; Curds *et al*., 1968; Fenchel and Harrison, 1976; Güde, 1979).

Competition for Nutrients. By competing with bacteria for dissolved organic nutrients and minerals protozoa may have a similar, although less direct, effect than that described above. However this kind of situation is unlikely to exist in many natural decomposer systems, polluted rivers or streams and sewage purification plants where continuous flow enrichment operates. It will occur in closed systems, although how much a fall in bacterial numbers is due to competition with protozoa for nutrients and how much is due to grazing by protozoa on the bacteria, is difficult to establish.

Population Studies of Protozoa in Decomposer Systems

The population of protozoa in a river or stream changes as a result of pollution by effluent discharges into the water mass. There is a considerable body of evidence accumulating to support this statement and a concensus of opinion that protozoa are good indicators of water quality (Henebry and Cairns, 1980b). Not so certain is how much the freshwater biologist can rely on particular indicator species of protozoa or whether the structure of entire communities is required.

Bick (1972), in his illustrated guide to protozoa, lists many indicator species. Two species of scuticociliates, the commensals *Heterocinetopsis unionidarum* and *Conchophthirus curtus*, were found to be more sensitive to pollution than their bivalve hosts (Antipa, 1977). *Heterocinetopsis*, an obligate commensal on the gills of *Anodonta*, disappears within two days in α-mesosaprobic conditions. Peritrich ciliates can be used to monitor concentrations of toxic chemicals such as mercuric chloride (Burbanck and Spoon, 1967).

Moderate organic pollution in a water course causes an increase in diversity and total numbers of protozoa (Henebry and Cairns, 1980b; Small, 1973), although toxic stress results in decreased diversity leaving only the more tolerant species (Cairns *et al*., 1980). The ciliate population in a small stream in Central Illinois was shown to increase below the outfall of a sewage treatment plant (Small, 1973). This was the situation at 300 metres below as compared with a similar distance upstream from the outfall. This clearly-defined 300 metre-below zone, containing a preponderance of bactivorous ciliates, was considered an extension of the sewage purification plant, a region of enhanced biological activity but with ciliated protozoa playing the major role. Analysing the food preferences of the ciliates found in this stream, Small found 22 species of algivores, 87 bactivores, 56 carnivores and 8 omnivores which formed into clearly-defined micro-biocoenoses, or micro-communities. In trying to categorize this particular Illinois stream, it probably corresponds with Kolkwitz and Marsson's α-mesosaprobic category on organic content and bacteria, but on Patrick's classification its wide variety of ciliates suggests a semi-healthy stream. As with all ecological studies generalizations are possible and useful but local conditions often create exceptions in particular ecosystems.

Further monitoring of streams and evaluation of entire communities of protozoa has involved the use of artificial substrates (Henebry and Cairns, 1980b). It was found that by submerging polyurethane foam units for colonization enhanced replicability in routine sampling was achieved (Cairns *et al.*, 1976a). Henebry and Cairns (1980b) confirmed that changes in the species – richness of entire protozoan communities can be measured easily and quickly on submerged units. Beyond one day no real increase in species resulted; at seven days the units tended to deteriorate as habitats through silt accumulation. Species – richness increased sharply below the point of discharge of an effluent, then gradually fell away (Figure 3.6). Applying similar techniques to sampling lentic lake water may require six weeks or more before colonization rate and extinction rate reach equilibrium (Cairns *et al.*, 1969). Interaction between colonizing species within the 'island' unit is probably more important than outside influences on the composition of the protozoan population (Cairns *et al.*, 1976b).

In the foregoing paragraphs it has been seen that a moderate input of organic effluent enhances variety in a stream by providing substrate directly for protozoan growth, or indirectly, through stimulating bacterial multiplication. More extreme organic pollution, particularly if high in carbohydrate and phosphorus content, causes the growth of amorphous masses of filamentous bacteria such as *Sphaerotilus natans*, the sewage 'fungus'. This provides a convenient substratum for those browsing and crawling protozoa which can tolerate near-anoxic conditions, a polysaprobic or α-mesosaprobic zone.

Another approach to the study of changes in populations of protozoa during decomposition of organic material is that of experimental decomposer systems as laboratory microcosms (Fenchel and Harrison, 1976). A clear succession, bacteria → flagellates → ciliates, could be traced as decomposition proceeded. Small flagellates, species of *Bodo*,

Figure 3.6: Increased Species Richness in Protozoan Communities in a Stream as a Result of Moderate Organic Pollution; ↓ indicates site of effluent discharge between Stations 2 and 3.

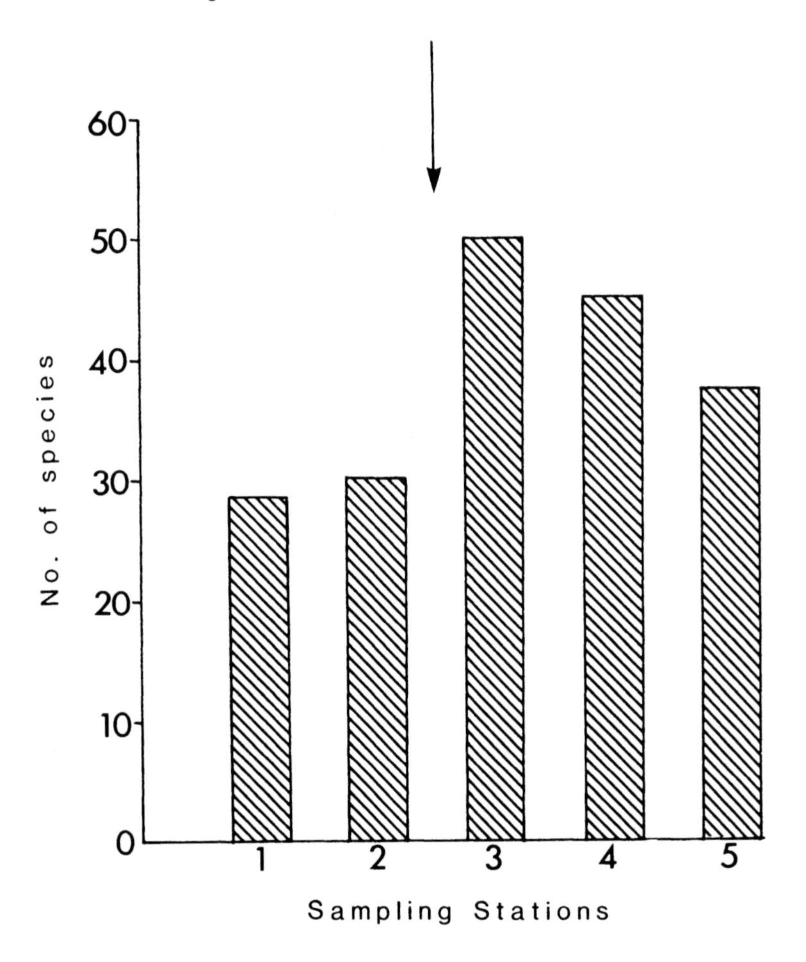

Source: Henebry and Cairns (1980b).

Monas, *Rhynchomonas* and some colourless euglenoid flagellates appeared after 20 hours and the population peaked at about 200 hours; it was followed by ciliates (*Euplotes*, *Holosticha*, *Uronema* and *Cyclidium*) at 50-100 hours and reached a population peak relatively more rapidly from 200 hours onwards. This climax population of ciliates could be expected to feed on any micro-organisms associated with the detrital mass, bacteria, unicellular algae or small flagellates. Such microbial decomposer systems give an indication of the trophic levels and the complexities of the inter-relationships in detritus-based food webs (Figure 3.7).

Figure 3.7: A Detritus-based Food Web, to Show the Nutritional Inter-relationships between Associated Micro-organisms. The species included are examples only.

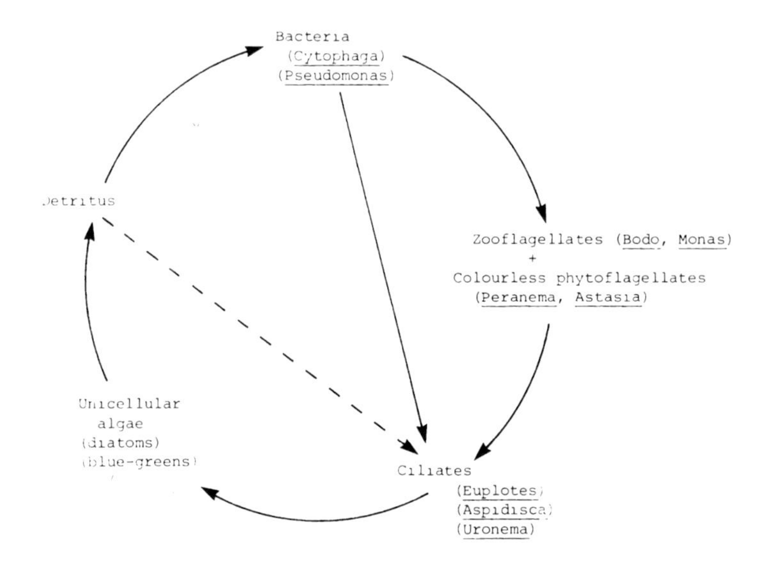

Bacterial shape is an important control on the grazing activities of small flagellates such as *Bodo* (Güde, 1979). Succession studies in continuous culture demonstrated that the grazing-resistant bacterium *Microcyclus* eventually outgrew the single-celled *Cytophaga* which was grazed upon by *Bodo*. The superior multiplication rate of *Cytophaga* enabled it to dominate the microbial population until *Bodo* was introduced (Figure 3.8). Grazing-resistance factors such as spiral and filamentous growth or floccing are protective strategies in *Microcyclus*-type bacteria, these can be lost after several generations of culture in the absence of protozoa.

Protozoa in Biological Purification Systems

The presence of protozoa in sewage purification systems, which prompted investigations into possible roles, has been recognised for many

Figure 3.8: Succession in Bacteria and the Flagellate *Bodo*, induced by grazing, grazing resistance and reproductive rate. ↓ indicates the times of adding *Cytophaga* sp. or *Bodo* sp.; ●—●, *Microcyclus* sp; ■—■, *Cytophaga* sp. Note the rapid decline in *Cytophaga* after the introduction of a selective bacterivore, *Bodo*.

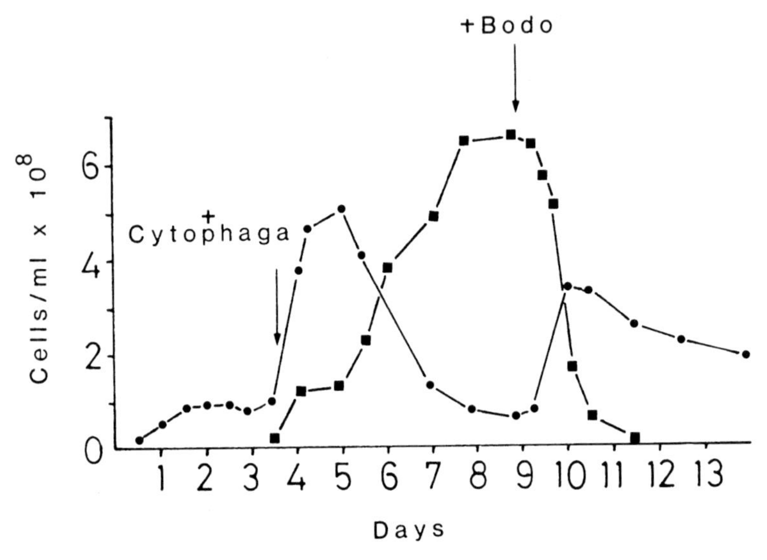

Source: Güde (1979), by permission.

years (Agersborg and Hatfield, 1929; Ardern and Lockett, 1928; Barker, 1942, 1943; Brink, 1967; Clay, 1964; Curds, 1975; Curds and Cockburn, 1970a, b; Hausman, 1923; Hawkes, 1963; Lackey, 1925; Morishita, 1976). The habitat provides a plentiful supply of most of the ingredients for prolific protozoan growth and a liquid very rich in organic matter and with a high bacterial population in consequence. The principal limiting factors then become (a) oxygen supply in certain types of sewage treatment plant and (b) the rate and volume of flow of liquid through the plant. Of the three types of sewage purification plant in common use, Imhoff tanks, percolating filters and activated sludge plants, the former are the most restrictive in terms of variety.

Imhoff Tanks. Some of the earliest work (Hausman, 1923; Lackey, 1925, 1932; Agersborg and Hatfield, 1929), centred round the protozoan fauna of Imhoff tanks, deep tanks of concentrated sewage liquor in which the high rate of bacterial decomposition made all except the top few centimetres anaerobic. Under these conditions the protozoan fauna is very restricted, including only those species which can survive with little or no oxygen, and which can tolerate relatively high levels of sulphides and ammonia. A fauna list of these obligate and facultative anaerobes includes a few ciliates, flagellates and sarcodines (Table 3.3).

Table 3.3: Anaerobic Protozoa of Imhoff Tanks.

Obligate Anaerobes	Facultative Anaerobes
Ciliates:	Ciliates:
Metopus es *Saprodinium putrinum* *Trimyema compressa*	–
Flagellates:	Flagellates:
Trepomonas agilis	*Bodo caudatus* *Dinomonas vorax*
Sarcodines:	Sarcodines:
–	*Vahlkampfia minuta* *V. guttula* *V. limax* *Hartmanella hyalina*

Percolating Filters. The principle of a percolating filter is well known. Waste water trickles over a suitable substrate of graded stones which quickly develops a film of microbial growth. This biofilm grows in an orderly sequence: zoogloeal bacteria collect around organic debris, followed by algae and fungi, and finally protozoa (Mack *et al.*, 1975). A list of species of protozoa found in percolating filters has been compiled by Curds (1975). It contains a preponderance of ciliates, 115 species, with 35 phytomastigophorean flagellates, 30 zoomastigophoreans, 29 rhizopodans and 7 actinopodans.

Once a mixed population of protozoa has developed, the quality of the effluent improves, particularly due to the activities of ciliated protozoa (Lloyd, 1945; Curds and Cockburn, 1970a). Certain species of ciliate are considered to be more important in the sewage treatment process than others. Table 3.4 lists nine species of ciliates in order of suggested importance, topped by the sessile peritrich species of *Opercularia*, *Carchesium* and *Vorticella* and the grazer, *Chilodonella uncinata,* all bacterivores.

Table 3.4: Ciliated Protozoa of Sewage Treatment Plants Listed in Order of Importance.

Percolating Filters	Activated-sludge Plants
Opercularia microdiscum	*Aspidisca costata*
Carchesium polypinum	*Vorticella convallaria*
Vorticella convallaria	*Vorticella microstoma*
Chilodonella uncinata	*Trachelophyllum pusillum*
Opercularia coarctata	*Opercularia coarctata*
Opercularia phryganeae	*Vorticella alba*
Vorticella striata	*Carchesium polypinum*
Aspidisca costata	*Euplotes moebiusi*
Cinetochilum margaritaceum	*Vorticella fromenteli*

Source: Curds and Cockburn (1970a)

The influence of these ciliates is a combination of direct consumption of sewage bacteria and indirect help in flocculating suspended finely-particulate detritus. The combined effect of removing bacteria by eating them and settling suspended detritus, gives a clear effluent which is the aim of the sewage purification process. The final settlement tanks in percolating filter systems often also support heavy growths of colonial peritrich ciliates whose combined ciliary activity may have a profound effect over a limited area. Ciliary activity, accompanied by secretion of mucopolysaccharide, from distinct mucocyst bodies in some ciliates, aggregates suspended particles by flocculation (Sugden and Lloyd, 1950; Curds, 1963). Experimentally it can be shown that a heavy growth of *Carchesium polypinum* will effect considerable clearing of a suspension of Indian ink (0.3% v/v) within 60 minutes, thus reproducing on a small scale the process in the final settlement tank at the sewage disposal plant (Figure 3.9).

Activated Sludge Plants. Protozoa are equally important in activated sludge plants, although the conditions in which they grow are considerably more rigorous. Because of the continual aeration and agitation, decomposition of organic matter proceeds faster and retention time in the sewage plant is shortened. The protozoa most favoured to maintain steady population numbers are those which either attach to the sludge surface (species of *Vorticella* and *Opercularia*) or those which browse on the flocs of sludge. In this category are the crawling hypotrichs, *Aspidisca* species, which Curds and Cockburn (1970a) found to be present in large numbers in 69 per cent of the samples

Figure 3.9: Clearing Suspended Particles by the Ciliary Action of *Carchesium*. (a) Ciliary currents create vortices of Indian ink particles which then settle in aggregated masses; (b) and (c) show the density of colonization of *Carchesium* used in (a).

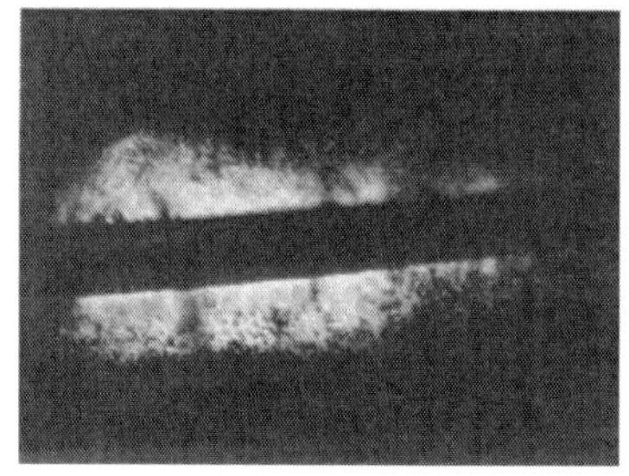

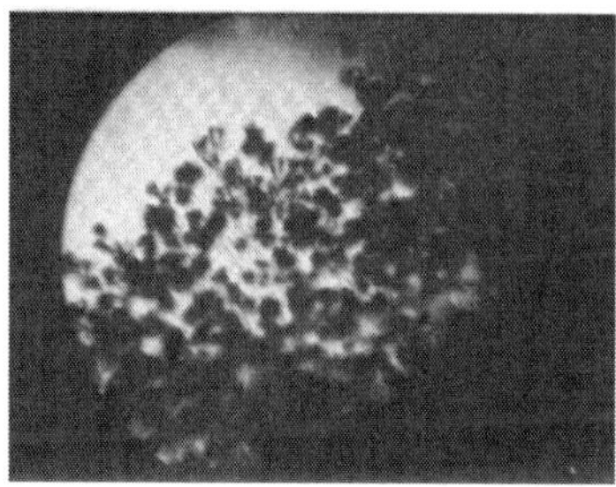

collected from activated sludge plants in widely scattered localities of England and Southern Scotland. The ecological strategy of these particular ciliates is reflected in their importance. A list of species, comparable to that in percolating filters, is seen in Table 3.4. The most obvious difference is the position of *Aspidisca costata* at the head of the list, but again peritrich ciliates predominate. In Curds' (1975) checklist there are 160 species of ciliates, 20 phytomastigophoreans, 16 zoomastigophoreans, 23 rhizopodans and 6 actinopodans. Information for this list was drawn from a variety of sources. In addition to those already quoted in this section should be added Brown (1965) and Schofield (1971), who made ecological studies of the micro-organisms in activated sludge plants.

Populations do exhibit some seasonal variation in activated sludge. Morishita (1976), comparing winter and summer populations, found 53 species in winter and 62 species in summer. A similar situation exists in percolating filters where Barker (1942) reported smaller and more erratic populations in winter. Listing activated sludge species in order of decreasing frequency, winter sludge contained *Monas*, *Aspidisca*, *Amoeba* and *Vorticella microstoma*; summer sludge contained *Epistylis*, *Amoeba*, *Vorticella* spp. and *Aspidisca*. It was interesting to note that here *Aspidisca* was never the dominant genus, although more nearly so in winter.

Active free-swimming ciliates such as *Paramecium* have a low frequency rating in activated sludge plants (Curds and Cockburn, 1970a). They are negatively geotactic and therefore are not retained on the settled sludge. Their relatively slow division rate does not reach the required 5.8 or more divisions a day which would be necessary for them to maintain populations in all except the very low flow rate sewage plants (Curds and Vandyke, 1966). The tendency is for free-swimming species to be washed out with the final effluent instead of being recycled with the activated sludge.

Small flagellates have been a somewhat neglected group in the past because of the difficulty of identification and their limited effect on the habitat. Recently Hanel (1979) investigated the ecology and systematics of 54 species of colourless flagellates found frequently in sewage treatment plants and listed 26 as being suitable indicators of water quality and reflecting the efficacy of the purification process (Table 3.5). In spite of Hanel's grouping and description of small flagellates in decomposer systems, their identification remains a problem to all but the specialist.

Looking for functions of protozoa in the biological treatment of

Table 3.5: Colourless Flagellates as Indicators of Water Quality

Polysaprobes	α-Mesosaprobes
Bodo edax	*Ancyromonas sigmoides*
Gyromonas sp.	*Anthophysa vegetans*
Helkesimastix faecicola	*Bicoeca lacustris*
Hexamitus inflatus	*Bicoeca petiolata*
Hexamitus pusillus	*Bodo angustus*
Mastigamoeba reptans	*Bodo saltans*
Monas ocellata	*Codonosiga botrytis*
Cicomonas mutabilis	*Cyathomonas truncata*
Polytoma uvella	*Monas guttula*
Tetramitus pyriformis	*Multicilia lacustris*
Trepomonas agilis	*Paraphysomonas vestita*
Trepomonas repans	*Rhynchomonas nasuta*
Trigonmonas compressa	
Urophagus rostratus	

Source: Hanel (1979).

sewage seems for the present to be out of favour. Perhaps also the trend towards more enclosed and controlled activated sludge plants will have the effect of reducing the number and variety of protozoa which find this habitat to their liking.

Associations with Other Animals

Associations between protozoa and other animals develop at many levels and may therefore be interpreted in their broadest sense to mean any grade of relationship, whether harmful or otherwise, which is of a permanent or semi-permanent nature. These relationships have been defined by various authors over the years and can be considered as grades of association which have no clearly-defined boundaries. The various terms of association are being used here consistent with common usage, following Whitfield (1979), to include *commensalism*, a relationship between two species of animal in which one benefits and the second is neither helped nor harmed, *symbiosis*, a relationship between two species in which both benefit and cannot exist independently, and finally, *parasitism*, a one-sided relationship in which one species lives in or on the body of another animal species, taking its nourishment and giving nothing in return, thus implying a detrimental association.

Successful Relationships with Other Animals

Protozoa are ideally suited for developing temporary or permanent associations with other animals. What special features do they possess which allow them to exploit a situation to its fullest advantage? They fall conveniently under four headings: small size, adaptability, capacity to produce cysts and great reproductive potential.

Small Size. Being small has advantages and disadvantages, but in this context an animal which is as small as an average protozoan, some 100 μm, can be housed very easily, often as a harmless ectocommensal using a larger animal as a means of transport. Small size also allows for ease of penetration into cavities in search of food and a congenial environment. One can speculate that many protozoa, particularly ciliates and amoebae of the gut, probably adopted this way of life as a result of accidental penetration when they were swallowed as cysts. Such protozoa have not changed much morphologically from their free-living counterparts.

Adaptability. This is an important factor in aiding the successful establishment of an infection after passage to a new host. Parasitic protozoa, particularly those which spend part of their lives in more than one host, have of necessity to adapt to very different environments. Several examples are found in those kinetoplastid flagellates (*Trypanosoma* and related genera) which spend part of their lives in vertebrate blood, which is oxygenated and rich in readily oxidisable sugars and, except in the lower vertebrates, is always warm. Their alternative habitat is the gut of an arthropod, low in oxygen and in soluble sugars and variable in temperature. The physiological and biochemical adjustments which are required of parasitic protozoa as a result of these differing life styles will be discussed in a later chapter.

Some species are particularly well adapted as endoparasites because of their capacity for antigenic variation which enables the parasite to produce different antigens sequentially to combat host antibodies (Vickerman, 1978).

Cysts and Other Means of Transmission. Many protozoa have the capacity to encyst when conditions become unfavourable, and this capacity assumes greater importance in the life cycles of many parasitic protozoa which have to solve the problem of passage from one host to another. Most parasites of the gut and body cavities must enter their host by oral injection and face the hazards of digestive enzymes in the foregut. Protected within cysts, the parasites can pass beyond the region of active digestion before emerging lower in the gut, but then must re-encyst before being shed in the faeces. This facility is exploited by parasitic amoebae such as *Entamoeba histolytica* which also uses the encysted state as a phase of multiplication.

One group of gut protozoa, ciliates of the ruminant forestomach, are an exception. They complete their life cycle without the need for cysts, for they are transmitted in saliva at rumination and their rumen environment is free from host digestive enzymes. The surplus population is digested when washed further through the gut as it has no protection.

Certain groups of parasitic protozoa, those of the blood particularly, have eliminated the need for cysts by the use of insect vectors which transmit infection when blood-sucking. The transmission of *Plasmodium* by anopheline mosquitoes, and of *Trypanosoma* by the tsetse fly, *Glossina*, have been recognized for many years.

Reproductive Potential. The capacity to exploit favourable situations by feeding, growing and reproducing over a very short time interval is clearly one of the more important features in considering protozoa in relation to other animals. *Plasmodium vivax* schizonts in human blood break up every 48 hours into 16-18 merozoites, each of which can infect a new red blood corpuscle in preparation for a repeat of the multiplication cycle and a bout of malarial fever. A protozoan which can increase its numbers rapidly by binary or multiple fission once inside a suitable host animal increases its chance of continuing the survival of that species. Passage from one host to another is a time of hazard, many parasites never finding the appropriate host; overproduction is therefore a necessary feature of all parasitic life cycles. The life cycle is punctuated by periods of rapid multiplication, often involving asexual alternating with sexual cycles. Phases of multiplication in the major parasitic group Apicomplexa involve multiple fissions, *schizogony* in the asexual phase of the life cycle and *sporogony* in the encysted zygote after fusion of the gametes. Blood-dwelling coccidians such as *Plasmodium* have evolved processes of rapid multiplication which they pursue with great efficiency, whilst exploiting three different habitats within their chosen hosts (Figure 3.10). The parasite invades its mammalian host as a sporozoite, passing briefly through the blood to exploit the resources of cells in the liver, the exo-erythrocytic stage, and then into the blood corpuscles, with phases of schi-

Figure 3.10: The Life Cycle of *Plasmodium*, Causative Organism of Human Malaria. Exo-erythrocytic and erythrocytic stages are found in man, followed by a sexual stage in the midgut of an anopheline mosquito. Infective sporozoites are transmitted through salivary gland excretion during blood-sucking. No part of the life cycle is passed outside a host.

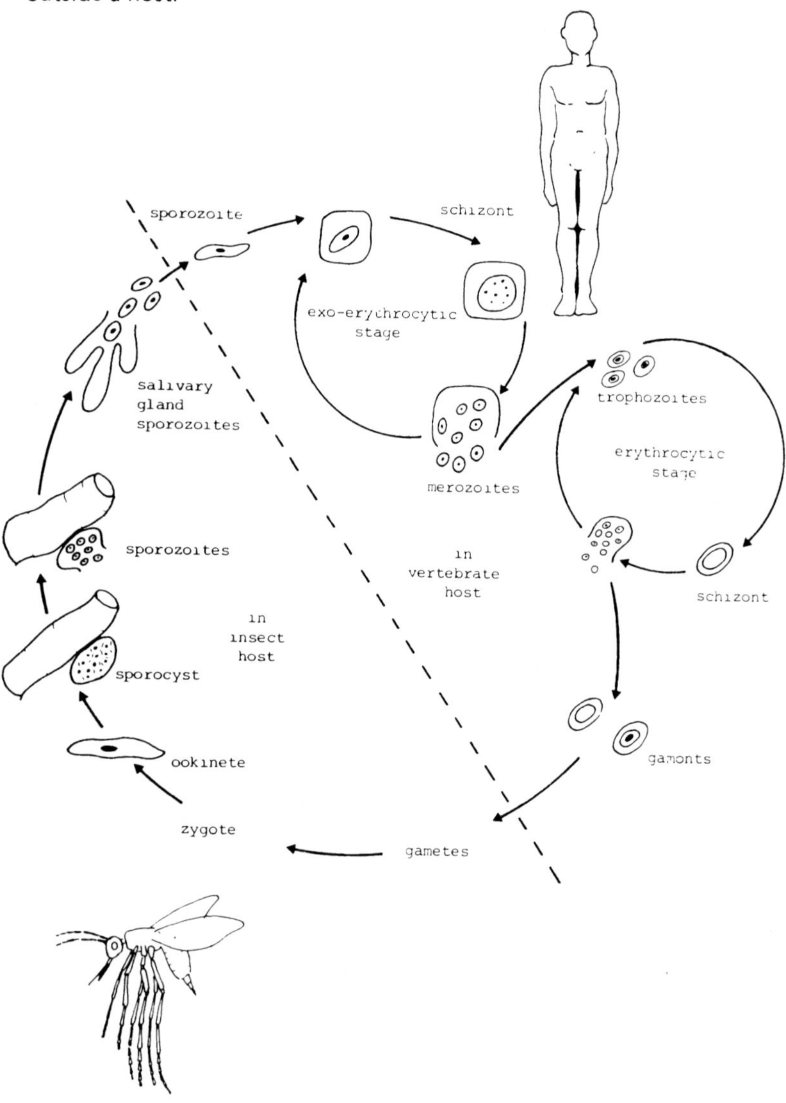

zogony in each host tissue. After successful transfer to the mosquito gut, further multiplication of the motile zygote, the ookinete, by sporogony produces large numbers of infective sporozoites. The greater the population the more certain is the chance of transfer to a new host.

Many apicomplexan parasites, particularly the gregarines, have abbreviated versions of the typical life cycle with fewer phases of multiplication. Because *Monocystis*, a parasite of the earthworm seminal vesicle, has no asexual multiplication in the trophozoite phase, the infection remains relatively harmless to its host. The real increase in numbers is during the encysted sexual phase, in gametogenesis followed by sporogony.

Commensalism

The many types of association which develop between protozoa and other animals may range from casual and temporary through to obligate and permanent. The loosest forms of association are those of commensalism, the ectocommensals which are often not excessively host-specific.

Ectocommensals. There are many examples, particularly among the ciliates, of epizoonts, protozoa merely adopting the practice of using other animals as a means of transport. Freshwater crustaceans, and to a lesser extent marine ones, collect growths of stalked peritrich ciliates, arborescent colonies sometimes positioned in such a way that they collect particulate food rejected or lost by the larger animal. A heavy growth of peritrichs may be no more than a temporary encumbrance to a swimming *Cyclops*, no parasitism being involved. Examples of epizoic protozoa (and epiphytic) are found among the groups of ciliates: (1) Chonotrichida – *Spirochona*, attaches to the branchiae on the thoracic limbs of freshwater *Gammarus*, *Stylochona* to marine *Gammarus*; (2) Peritrichida – *Epistylis* and *Carchesium* to various fresh and salt water crustaceans; (3) Suctoria – *Ephelota* on hydrozoan colonies such as *Obelia* and *Dendrocometes* on *Gammarus*. The associations in which no adhesive attachment is involved, but the connection is nevertheless as clearly established, are more difficult to understand because of their selectivity. Such relationships exist in *Kerona polyporum* crawling over the surface of *Hydra* and mopping up bacteria and detritus which fall on the sessile polyp, although this ciliate will also ingest *Hydra* tissue. The frequency with which *Kerona* is found on *Hydra* suggests that this is an established partnership. The relationship between various thigmotactic ciliates and bivalve molluscs, (e.g. *Conchophthirus curtus* in *Anodonta grandis*), where the mantle cavitics and gill filaments provide a very suitable hunting ground, also illustrates this level of association.

Endocommensals. These live inside other animals, ingesting food which is no longer of interest or value to the host animal, which means that the lower end of the alimentary canal, beyond the major site of absorption, is one such habitat. *Opalina* commonly lives in the rectum of frogs and toads in large numbers as a harmless commensal. There may be several hundred active *Opalina*, even though each is larger than an average protozoan. Although the small *Entamoeba coli* lives in the gut of its vertebrate host and ingests food, its presence will pass unnoticed as it does not invade the epithelial cells of the gut mucosa unlike the closely-related *E. histolytica* which causes extensive damage to the lining of the gut.

Symbiosis

Protozoa which earn their living, giving benefit to the host animal as well as receiving shelter and nutrients, are involved in a symbiotic association. Among this group are the well-known examples of flagellates of termites and woodroaches whose presence in the hindgut is essential for the digestion of plant tissues ingested by the insects. Whether this cellulolytic role is performed directly by the flagellates or indirectly through bacterial activity has not been firmly established. Certainly termites deprived of their protozoan fauna will starve to death because their diet of lignified material is otherwise indigestible.

Many herbivorous mammals, in particular ruminating artiodactyls, rely on their rumen micro-organisms (ciliates, flagellates and bacteria) to digest the plant material which is a major part of their diet. Because of the obvious economic importance of understanding ruminant digestion the activities of these micro-organisms have been studied in detail. The metabolic activities of the rumen microflora are of considerable interest but they are not part of this treatise. The rumen protozoa are discussed in detail in Chapter 8. All the protozoa associated metabolically with ruminants belong to the ciliate groups Trichostomatida and Entodiniomorphida and such is their adaptation to life in the rumen that they are incapable of prolonged life outside. However, the trichostomatid representatives, *Isotricha* and *Dasytricha*, look little different from their free-living counterparts, i.e. they have a contractile vacuole, although they do not need to use it. The entodiniomorphids have morphological specialisms which are not found elsewhere, in particular the 'skeletal' plates whose presence is difficult to justify as supporting structures in organisms which have such a tough pellicle that this pellicle remains intact and identifiable for many hours after death.

Another type of symbiotic association exists in which the protozoon can be considered as playing the role of host animal, and the other partner is an autotrophic unicellular alga. Ciliates such as *Paramecium bursaria* have established a nutritional partnership with the green alga, *Chlorella*, which will be discussed further in Chapter 8. These symbiotic associations are not restricted to ciliates, for some naked and shelled amoebae also utilize the nutritional expertise of algae to augment their own heterotrophic syntheses.

Parasitism

Parasitic protozoa have occupied a central role in protozoology for many years. Interest is gradually moving away from general aspects of biology and life cycles to the more specific problems of immunology and surface membrane behaviour. The nature of the interaction between host cells and the parasite, the mechanism of penetration into host cells and the nature of the protective mechanisms which parasites adopt in different hosts are some of the rapidly expanding fields in modern protozoology. The specialist nature of this research makes detail inappropriate here.

The capacity for antigenic variation which has already been mentioned is an important aspect of the host-parasite relationship. Trypanosomes, in the metacyclic form in the salivary gland of their tsetse fly host, prepare for transfer to the blood of their mammalian host by acquiring a surface coat of a variable antigenic type. Recent research has shown that there is not one basic variable antigenic type: there are several (Vickerman *et al.*, 1980). *Trypanosoma* has developed a very efficient protective mechanism. Thus the problems for a mammalian host in producing effective antibodies against *Trypanosoma* infections are numerous. The possibility of developing anti-sleeping sickness vaccines is also unlikely at present (Vickerman, 1978).

Many parasites live within the cells of their host. Amongst the Apicomplexa, coccidians and piroplasms are typically intracellular and enter by actively inducing the host cell to phagocytose them. This involves an interesting surface reaction between the membranes of the parasite and host. The events accompanying the entry of a malarial parasite (*Plasmodium*) into a mammalian blood cell have been reviewed by Wilson (1982). The process involves a series of steps: *recognition* of an appropriate cell; *attachment* to the cell membrane if the parasite is correctly orientated; and *junction formation* culminating in *entry* into the red blood cell. Nutritional interactions between parasites and their hosts will be discussed in the systematic sections in later chapters.

In Summary

Inter-relationships and interactions are accepted phenomena of animal communities. In discussing interspecific associations, however, it must be recognized that definitions such as are used in this section cannot cover all instances of associations. But broad categories are necessary, and commensalism, symbiosis and parasitism create a framework for use in the following chapters. The view taken by Lee (1980a) is certainly correct: 'in analysing feeding mechanisms and nutritional requirements one must avoid static concepts of habitats and niches'; but without some concept of habitats and niches there is no starting point. The last two chapters have examined a wide variety of habitats in which protozoa are found, together with likely food sources. The logical next step is to consider in general terms how nutrients are incorporated into protozoan cytoplasm.

4 AUTOTROPHIC NUTRITION

In the preceding two chapters an attempt was made to relate the feeding habits of protozoa in selected habitats to the kinds of food available to them and to the competitive pressures from other organisms sharing these same niches. The logical pursuit of this theme is next to consider how the various groups of protozoa utilize the food which they have acquired. While recognizing that most protozoa are animal-like in their nutritional processes, provision must be made to include the range of flagellates, some of which are pigmented and light-dependent and others which, although pigmented, have the capacity to switch reversibly to alternative methods of nutrition in the absence of light.

Classifying Nutrition

Any schemes for classifying living organisms into nutritional types 'depend on the organism's sources of energy, carbon and reducing equivalents' (Hamilton, in Lynch and Poole, 1979). If emphasis is placed on carbon source then grouping organisms as *autotrophs* which obtain their cell carbon from carbon dioxide or other one-carbon compounds and *heterotrophs* which require pre-synthesized organic compounds, remains an acceptable classification. This is perhaps not comprehensive enough for modern thinking. Alternatives produce a bewildering variety of terms. The nature of the energy source divides organisms into *phototrophs* which utilize energy from the sun and *chemotrophs* which obtain energy by the oxidation of organic or inorganic compounds. The source of reducing equivalents for cell synthesis labels organisms as *lithotrophs* when utilizing inorganic and *organotrophs* when using organic sources.

In considering only protozoan nutrition it is possible to limit the wide-ranging scheme just described. Three compound terms have been selected as the most appropriate and descriptive.

(1) Photoautotroph. Many flagellates are autotrophs which obtain their energy from the sun (phototrophy), and thus can be called *photoautotrophs* without sacrificing any accuracy of terminology. This process of trapping light energy for the synthesis of glucose units and other cell

carbon compounds from carbon dioxide is *photosynthesis*. It may be considered by many biologists to be of only peripheral interest in protozoan nutrition and yet a large section of the Mastigophora rely on light as a primary source of energy.

2) Photoheterotroph. A number of flagellates, although phototrophic in energy requirements, cannot use carbon dioxide for cell synthesis and must have organic carbon compounds. These may be appropriately thought of as photoheterotrophs.

(3) Chemoheterotroph. Organisms which require chemical energy and organic carbon sources are the chemoheterotrophs. This is the largest nutritional category which encompasses most protozoa and all other animals.

The nutritional versatility of euglenoid flagellates particularly, which has already been mentioned, allows these flagellates to change from photoautotrophy to chemoheterotrophy in the dark if suitable organic compounds are available. It is important to point out at this stage that although many euglenoid flagellates are essentially photoautotrophic in light, they are incapable of growing without at least an inorganic nitrogen source and certain vitamins. The kind of facultative switch found in euglenoids and many other pigmented flagellates makes the allocation of protozoa to nutritional categories a far from simple exercise. A more detailed consideration of this problem is to be found in Chapter 9.

Another form of nutritional versatility involving photoautotrophy appears in those protozoa (ciliates and amoebae mainly) which contain populations of actively photosynthesizing organisms. These endosymbionts, whose existence has already been noted in *Paramecium bursaria*, pursue their photosynthetic activities to the benefit of the ciliate and the symbiont (see later).

The Process of Photosynthesis

In simple terms photosynthesis, as found in photoautotrophic organisms such as *Euglena*, can be expressed by the equation:

$$CO_2 + H_2O \xrightarrow{\text{light}} (CH_2O) + O_2$$

but this does not give a comprehensive picture of the nature of the reaction. Photosynthesis is an energy-generating process where light energy is converted into chemical energy for the metabolic activities of the cell. Units of light energy, or photons, emitted by the sun are absorbed on to photosynthetic pigments, which in phytoflagellates, eukaryote algae and green plants lie in the lamellar bands of the chloroplasts. Chemically these pigments are usually either chlorophylls, carotenoids or phycobilins. The important pigment in' relation to light absorption in flagellates such as *Euglena* is chlorophyll a, the molecules of which are the reaction centres on the lamellae of the chloroplasts. Other chlorophylls and carotenoids are accessory pigments which act as light gatherers over different wavelengths, transferring their light energy to the reaction centres on chlorophyll a. Chlorophyll b, present in the green algal groups of euglenoids, volvocids and chloromonads, functions as an accessory pigment passing its light energy to chlorophyll a (Duysens and Amesz, 1962). Carotenes and phycobilins also act as light gatherers over different wavelengths, transferring their light energy to the reaction centres on chlorophyll a (Govindjee and Braun, 1974). Carotene also has an additional role in protecting chlorophyll from photodestruction, although the mechanics of this activity are not yet fully understood.

Photosynthesis is conveniently divided into two stages, the light reaction and the dark reaction. The *light reaction*, in which chlorophyll a plays the key role in harnessing energy released by the sun, is a non-cyclic phosphorylation process in green flagellates such as *Euglena*. It functions for the production of energy-rich ATP (adenosine triphospate) and NADPH (reduced nicotinamide adenine dinucleotide phosphate) required in the dark reaction and elsewhere in the cell metabolic reactions (Figure 4.1). In conditions of very low light intensity extra supplies of ATP and NADPH are required to be produced by cyclic phosphorylation in some algae (Raven, 1970, 1971; Tanner *et al*., 1969). The *dark reaction*, or Calvin cycle, is a cyclic process which incorporates carbon dioxide into glucose units which are the starting point for various anabolic and catabolic conversions essential for cell functioning (Figure 4.2).

The Light Reaction

Light energy is transported to reaction centres, large groups of chlorophyll a molecules within the lamellae of the chloroplasts, from which the chlorophyll releases a flow of electrons towards a strong electron acceptor. Indeed the whole light reaction can be thought of as two transport systems passing electrons from one electron acceptor to

Figure 4.1: The Light Reaction. ATP and NADPH are produced by a non-cyclic photophosphorylation in pigmented flagellates such as *Euglena*. Q and Z denote initial energy acceptors; plastocyanin, a copper-containing intermediate, is not confirmed in *Euglena*.

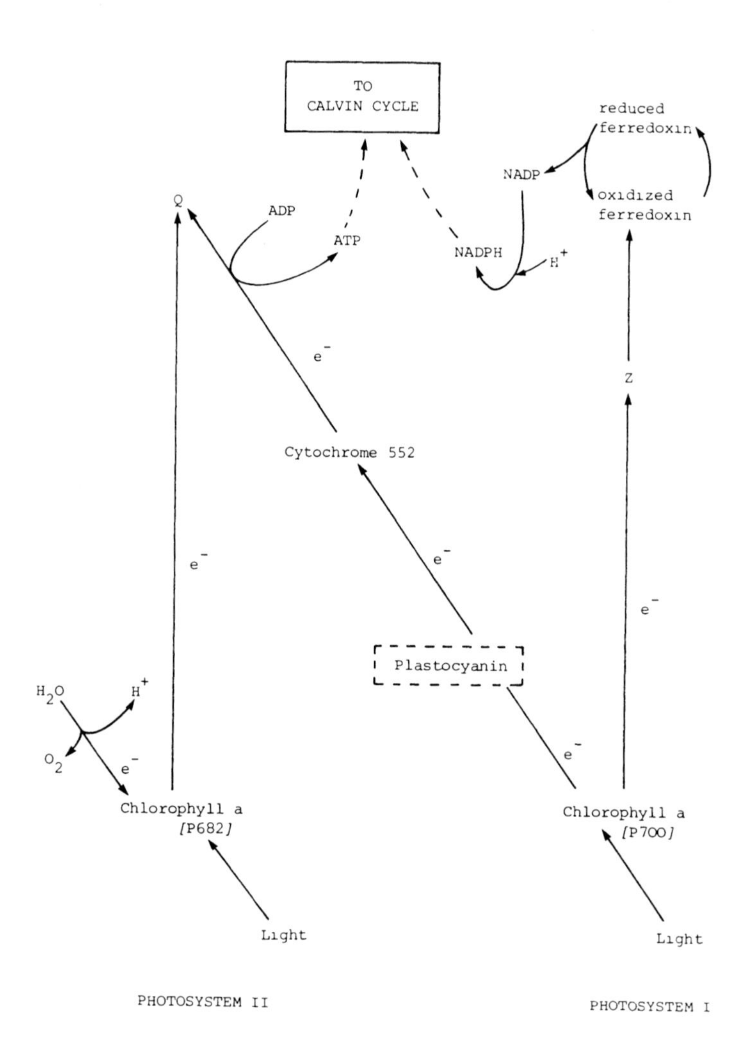

Source: Modified from Hill and Bendall (1960).

Figure 4.2: The Dark Reaction (Calvin Cycle). Adenosine triphosphate (ATP) and reduced nicotinamide adenine dinucleotide phosphate (NADPH) produced in the Light Reaction are incorporated into the Calvin Cycle where energy is required for polymerization.

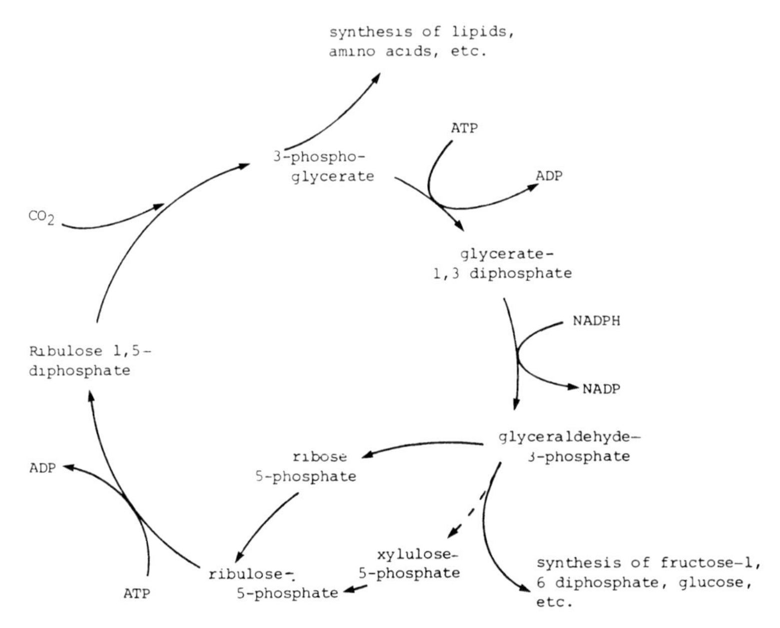

another until they reach their final destination. The two systems contain chlorophyll a molecules of differing absorption spectra drawn from a range of chlorophylls which are spectroscopically distinguishable by the location of their absorption maxima, from 660 nm with intermediates to 705 nm (French, 1971). *Euglena,* for instance, contains 670 and 680 nm absorption forms of chlorophyll a in light conditions, but develops 695 nm chlorophyll a in low light intensities (Evans, 1968).

Energy is absorbed by these various chlorophyll a and other light-gathering pigment molecules before being transferred to reaction centre chlorophyll a, absorbing at 700 nm (designated P700) in Photosystem I and at 682 nm (designated P682) in Photosystem II (Figure 4.1). For a more detailed account of light absorption in photosynthesis reference should be made to Govindjee and Braun (1974).

Photosystem I. In System I, as the electrons are drawn towards the initial electron acceptor, a hypothetical molecule designated Z, the chlorophyll molecules are left in an oxidized state ready to accept electrons from another donor. Z quickly transfers its electrons to chloroplast ferredoxin which in the reduced state passes them on to NADP. The addition of electrons to NADP allows it to incorporate hydrogen ions from the medium to form NADPH. NADPH retains the electrons it has acquired and itself passes to the Calvin cycle where it is involved in the synthesis of carbohydrate. Thus there is a net loss of electrons from Photosystem I and this deficit is compensated for by Photosystem II.

Photosystem II. This system operates through chlorophyll a (P682) reaction centres. Electrons pass to the initial electron acceptor, designated Q, which releases them through a chain of electron acceptors, to fill the deficit in Photosystem I (Figure 4.1).

In all photosynthetic systems, cytochromes act as electron acceptors at this linking point between the two systems. The important cytochrome in *Euglena* is Cytochrome 552, a single c-type cytochrome not containing the f-type component found in higher plants (Evans, 1968). Cytochrome 552 is absent in dark-grown *Euglena* but reappears in the light, in conjunction with chlorophyll as the chloroplasts develop. The presence of copper-containing plastocyanin, which in higher plants is the next step in the energy transfer between Systems I and II, has not yet been confirmed in *Euglena*.

The transport of electrons from Q to the beginning of Photosystem I involves a sequential drop in potential, resulting in spare electrons being made available to synthesize ATP from ADP. This ATP is incorporated into the reactions of the Calvin cycle and other processes important to the cell. The net result of these transfers creates an electron deficit in Photosystem II which is filled by a special characteristic of chlorophyll P682. This molecule is able to draw extra electrons by the splitting of water molecules into free oxygen and hydrogen ions by a process involving manganese ions lodged in the reaction centres. This rather simplified overview of the light-mediated reactions of photosynthesis shows how light energy has been incorporated into ATP and NADPH synthesis and how a further input of energy has been acquired indirectly through the splitting of water molecules. This partly validates the simple equation presented in the introduction of photosynthesis, that carbon dioxide and water combine to form carbohydrate.

The Dark Reaction

As the name indicates, this part of the photosynthetic process has no need of light to drive the chemical reactions involved. It is the carbon reduction cycle, or Calvin cycle, so named because Calvin and his associates over a period of years elucidated the mechanism by which carbon dioxide is incorporated into organic compounds (Bassham & Calvin, 1957). Figure 4.2 shows the cyclic nature of the process, but in the interests of simplicity no attempt has been made to balance the molecular equations. It does show the stages at which the products of the Light Reaction, ATP and NADPH, are incorporated.

Briefly, carbon dioxide combines with ribulose-1, 5-diphosphate to form 3-phosphoglycerate. This is further phosphorylated by ATP (from the Light Reaction) to glycerate-1, 3-diphosphate, which is then reduced by NADPH to glyceraldehyde-3-phosphate. One of the main metabolic pathways leading from the carbon reduction cycle is from here: glyceraldehyde-3-phosphate provides the starting point for the synthesis of hexose phosphates, which are required for the production of complex storage carbohydrates. The other principal metabolic pathway diverges from 3-phosphoglycerate towards the synthesis of lipids and amino acids. Ribulose-5-phosphate is regenerated by various routes to complete the Calvin cycle. The utilization of hexoses, lipids and amino acids are not processes unique to photosynthetic organisms and therefore the mechanics will not be considered at this point.

Confirmation that the Calvin cycle does indeed operate in green flagellates, including *Euglena*, has come from Latzko and Gibbs (1969) and Smillie (1968), who have shown that all the necessary enzymes are present in the chloroplasts. Also, isolated chloroplasts of *Euglena* are capable of fixing carbon dioxide at 50-100 per cent of the *in vivo* rate (Forsee and Kahn, 1972).

Organelles of Photosynthesis

Various organelles and structures within the photoautotrophic organism are directly involved in the process of photosynthesis and will be included in this section. Other cytoplasmic organelles, not specific to this process, do not justify mention here.

Chloroplasts are the principal organelles of photosynthesis, for they contain chlorophylls and other light-gathering pigments housed in lamellated thylakoids. A means of responding to light intensity, for protection against excessively intense light or for promoting movement towards stronger light, are equally important to photoautotrophic organisms. This stimulus-response system found in green algae and flagellates includes a photoreceptor, a sensory transduction system and an effector (Melkonian and Robenek, 1979).

Chloroplasts

In algae and flagellate protozoa chloroplasts may vary considerably in shape and number although their basic structure is similar. Ultrastructural studies have shown that chloroplasts consist of layered lamellae with each lamella containing a number of thylakoid units. Some of the early work on *Chlamydomonas* chloroplasts by Sagar and Palade (1957), then by Gibbs (1960, 1962b) on *Euglena* and other algae, led the way towards understanding chloroplast structure. The lamellated discs described by them were later named 'thylakoids' by Menke (1962).

Thylakoids are embedded in a proteinaceous stroma which is enclosed in a chloroplast envelope, normally of two unit membranes. Exceptions to this are the triple membrane envelopes of euglenoids and dinoflagellates, formed by fusion of an envelope from the chloroplast endoplasmic reticulum and extensions of tubular endoplasmic reticulum from the nuclear envelope (Leedale, 1967, and Figure 4.3).

When looking at chloroplast structure in other photosynthetic organisms, clear phylogenetic differences emerge. Thylakoids in prokaryotic cyanobacteria and other bacteria lie free in the body of the cell instead of being enclosed in a chloroplast envelope, which suggests a more primitive condition. Higher plants, on the other hand, although possessing membrane-bound chloroplasts, show more internal specialization. Disc-like thylakoids are closely stacked into grana separated by rather loose stroma lamellae which give the plant chloroplast a banded appearance not seen in lower organisms.

Thylakoids. These are the chlorophyll-containing units of the chloroplast and, as such, are sites for the photochemical reactions of photosynthesis. The structure of a *Euglena* thylakoid can be seen in Figure 4.3. The stacking of thylakoids into multiples ranging from two to six provides some basis for identification at order level although variability in the Euglenida means that this character must be used with caution. Chloroplasts of *Euglena gracilis* show thylakoids normally stacked in twos and threes with the outer lamellar membrane thin (5-6 nm) and the intervening ones thicker (10-12 nm). This appearance of thicker membranes is due to pairs of thin membranes from adjacent thylakoids being very closely adpressed (Gibbs, 1960). The chloroplasts of cryptomonads contain thylakoids loosely packed in pairs and in chry-

Figure 4.3: Chloroplast Structure of *Euglena gracilis*, Showing Two- and Three-thylakoid Lamellae and the Three-membraned Chloroplast Envelope.

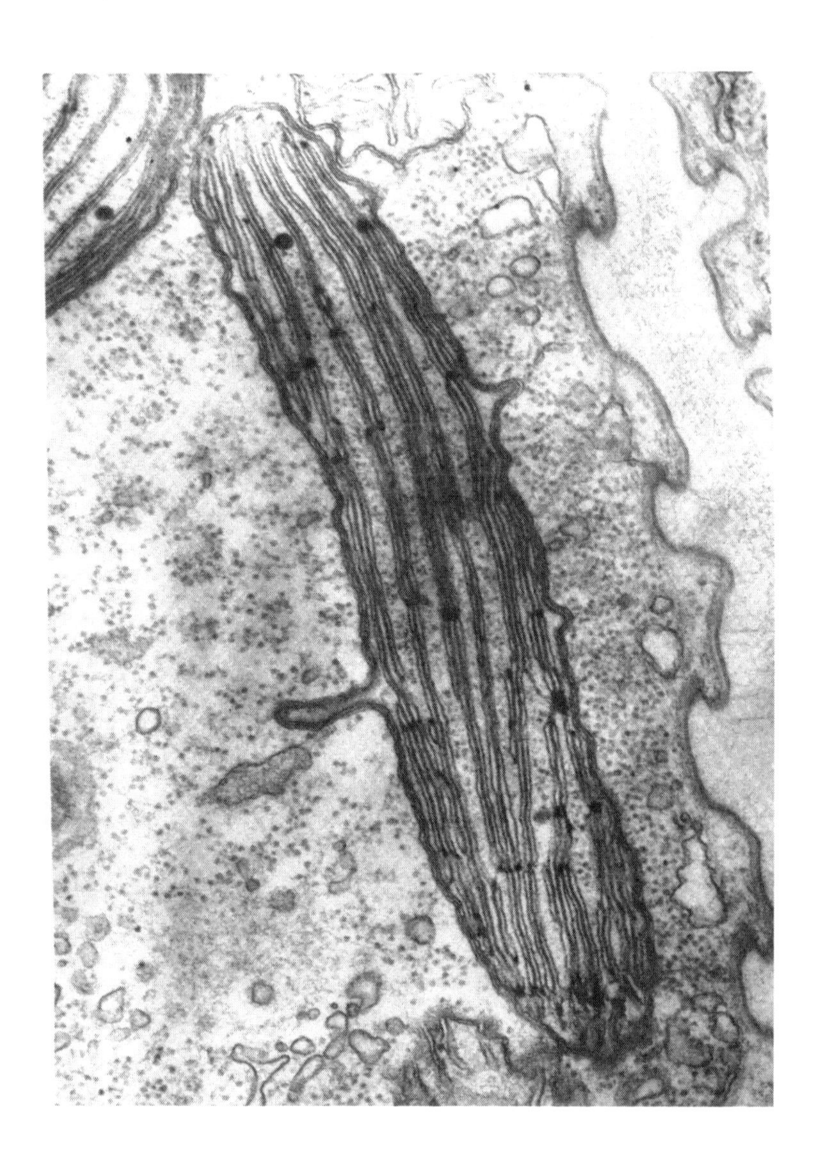

Electron micrograph by G.F. Leedale, x 80,000.

somonads they are arranged in threes (Gibbs, 1962b).

A five category classification of thylakoids in chloroplasts according to structure and distribution within flagellate and algal groups may be found in Bisalputra (1974), from which Figure 4.4 has been constructed.

Figure 4.4: Chloroplast Envelopes and Thylakoid Arrangements in Flagellate and Algal Groups. (a) Rhodophyta, single thylakoids not banded; (b) Cryptomonadida, paired separated thylakoids; (c) Dinoflagellida and Chrysomonadida, thylakoids in bands of three but separated; (d) Euglenida, thylakoids fused in twos or threes; (e) Volvocida and Prasinomonadida, two to six fused thylakoids.

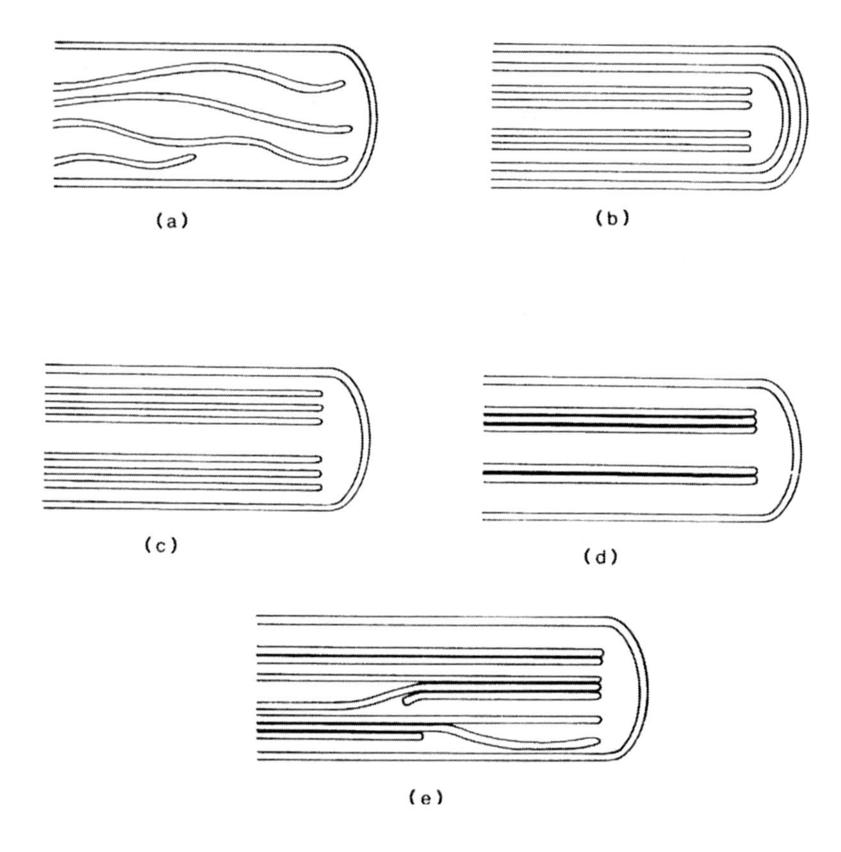

Pyrenoids. A distinctive character of the chloroplasts of many flagellates and algae is the presence of pyrenoids, not found in green plants.

A pyrenoid is a protein-rich organelle either embedded in the chloroplast, from which it can be distinguished by its denser matrix, or projecting from it as a knob or stalked structure. Isolated chloroplast lamellae often traverse the pyrenoid at widely-spaced intervals.

The pyrenoid functions as a site for the storage of enzyme protein for the chloroplast and may disappear temporarily in non-photosynthesizing forms (Gibbs, 1960). Reserve polysaccharide (starch) may also be present, laid down around the pyrenoid as a sheath or membrane in the volvocids and prasinomonads. Amongst the other flagellate orders carbohydrate reserves are outside the chloroplast membrane, as starch platelets adjacent to the projecting pyrenoid in dinoflagellates (Dodge and Crawford, 1971), or as isolated bodies of paramylon in the euglenoids (Gibbs, 1962a; Leedale, 1968) and of leucosin in chrysomonads (Gibbs, 1962b). The variable nature of carbohydrate reserves in phytoflagellates is an indication that the relationship between these orders is not particularly close. Characterization of carbohydrate reserve within the flagellate orders is further detailed in Table 7.1.

Pigments

The functioning of chlorophyll molecules as light reaction centres is a clearly established fact, as is the location of this pigment in the chloroplasts. Although chlorophyll a is the operative pigment in photosynthesis, flagellates contain a variety of accessory pigments. Other forms of chlorophyll and various red, yellow and brown carotenoids (carotenes and oxycarotenes or xanthophylls) and red and blue phycobilins give characteristic colour to the organisms. Chemically, chlorophyll is a complex molecule with a ring structure similar to that of the haemes, but containing magnesium chelated into the central ring nucleus of the molecule instead of iron. Carotenoids are essentially long unsaturated hydrocarbons with a series of attached methyl groups. Table 4.1 shows the distribution of the principal pigments in some flagellate orders.

It can be seen from Table 4.1 that β-carotene is uniformly present, but the different forms of chlorophyll give some basis for grouping according to evolutionary trends, which is also reflected in the number of xanthophylls identified. In possessing chlorophylls a and b the orders Euglenida, Volvocida and Chloromonadida separate as the 'green' line and those orders with chlorophylls a and c, the 'brown' line. Stewart (1974) provides a more detailed analysis of the distribution of photosynthetic pigments.

Table 4.1: Distribution of Photosynthetic Pigments in some Flagellate Orders.

Order	Chlorophyll		Carotene	Xanthophylls**	Phycobilins**
Chrysomonadida	a	c	β	3	–
Cryptomonadida	a	c	*α β	3	2
Dinoflagellida	a	c	β	3	–
Prymnesiida	a	c	β	3	–
Euglenida	a	b	β	many	–
Volvocida	a	b	β	many	–
Chloromonadida	a	b	β	?	–

Notes:
* More α than β.
** An indication of the range present in quantity.

Source: Adapted from Stewart (1974).

Eyespots and Orientation

To ensure that a suitable light intensity is available for photosynthesis, pigmented flagellates have special structures associated with orientation. The red or yellow eyespot usually located near the base of a flagellum consists of a cluster of lipid globules which are thought to function in light absorption. The lipid globules are arranged in two or more rows as a shallow concave disc lying within the chloroplast membrane and attached to the plasmalemma by pin-like projections 100 nm long (Melkonian and Robenek, 1979). An exception to this is found in the euglenoid flagellates where the eyespot is never within the chloroplast (Leedale, 1967).

Response to light involves a photoreceptor which can induce movement either towards or away from light, depending on its intensity. The exact location of photoreceptors in pigmented flagellates has not been established although it seems that in the green flagellate *Tetraselmis* (Prasinomonadida) the receptor corresponds to a region of electron-dense particles in the outer chloroplast membrane adjacent to the eyespot (Melkonian and Robenek, 1979).

The phototactic movement is induced by an electric potential which passes between the photoreceptor and the motor apparatus, the plasmalemma and the flagellar membrane (Litvin *et al.*, 1978; Marbach and Mayer, 1971). Efficient transmission of the electric potential is facilitated by the gap between the outer chloroplast membrane and the plasmalemma, which at this point is fixed at 25 nm and is devoid of cytoplasmic material (Cavalier-Smith, 1978; Melkonian and Robenek, 1979).

In *Euglena gracilis* the light receptor is a swelling at the base of one of the flagella (Gibbs, 1960; Leedale, 1967). In *Euglena*, as in other pigmented euglenoids which have more than one flagellum, this paraflagellar swelling is located near the base of the longer flagellum. Location of light-sensitive structures near the base of the longer flagellum which provides the motive force in locomotion, would seem to emphasize their importance in orientating the organism at the appropriate level in the water.

The need for light as an energy source in those organisms which rely entirely or mainly on photoautotrophic nutrition is a primary influence in distribution. The preponderance of pigmented flagellates over colourless forms in plankton samples taken in marine and freshwater habitats was discussed more fully in Chapter 2. In shallower bodies of water, small ponds and pools, depth adjustment is relatively ineffective as the light intensity will hardly vary between the surface and the bottom. Nevertheless, even obligate photoautotrophs will avoid subjecting themselves to over-illumination when the sun is directly overhead.

In Summary

This chapter has described photoautotrophy, more commonly called photosynthesis, as a two-stage process consisting of the Light Reaction and the Dark Reaction or Calvin Cycle. It has outlined the biochemical pathways which operate at both stages when obligate and facultative photoautotrophic flagellates utilize the sun as an extraneous energy source. For the sake of simplicity it has been assumed that carbon dioxide is the principal carbon source as in green plants, when in reality most flagellates have more complex requirements. Chapter 9 discusses the nutritional needs of flagellates in more detail.

The structure of chloroplasts and the distribution of pigments in relation to flagellate groups have been considered. A brief description of the eyespot and its involvement in photo-orientation have been included.

5 HETEROTROPHIC FEEDING

Much has been written on the evolution of metabolism and it would be superfluous to consider more than the barest introduction in this context. Most biologists accept that a simple form of heterotrophy was the means by which the earliest forms of life, the primitive prokaryotes, obtained energy to grow and replicate, utilizing preformed nucleotides and amino acids. That they did so in an anaerobic atmosphere is also accepted. As these early microbes proliferated one can visualize the situation whereby their metabolic requirements exceeded the supply of organic substrates produced abiotically in the 'primeval soup'. Some of these primitive organisms (early chemoautotrophs) then evolved the capacity to fix atmospheric carbon dioxide. Evidence that phosphoglyceric acid, phosphoglyceraldehyde and ribulose, all products of the Dark Reaction (see Chapter 4), could have been produced by primitive microbes, comes from the existence of similar metabolic pathways in anaerobic prokaryotes alive today. These carbon dioxide fixation pathways resulted in the accumulation of oxygen in the Precambrian atmosphere, preparing the way for aerobic heterotrophs to flourish, from prokaryotes through to eukaryotic micro-organisms.

Evolution of Nutritional Strategies

Protozoa have certainly evolved from aerobic prokaryote heterotrophs, although whether photoautotrophs were an essential link in the chain or whether they formed an alternative evolutionary pathway is an open question. Several views of the evolution of nutritional strategies should be considered. The traditional, and for a long time, the only view, was that the prokaryote photosynthesizers gave rise directly to eukaryote photosynthesizers (phytoflagellates). Some of this group became chemoheterotrophs (zooflagellates) by loss of pigment, leading finally to other protozoan groups and higher animals (Figure 5.1a). This seemed to offer a reasonable explanation for the existence of mixed nutrition groups (euglenids and cryptomonads) amongst the phytoflagellates, where pigmented and colourless forms occur side by side.

A second view presents the eukaryote photosynthesizers as a relatively recent development requiring the evolution of complex organelles

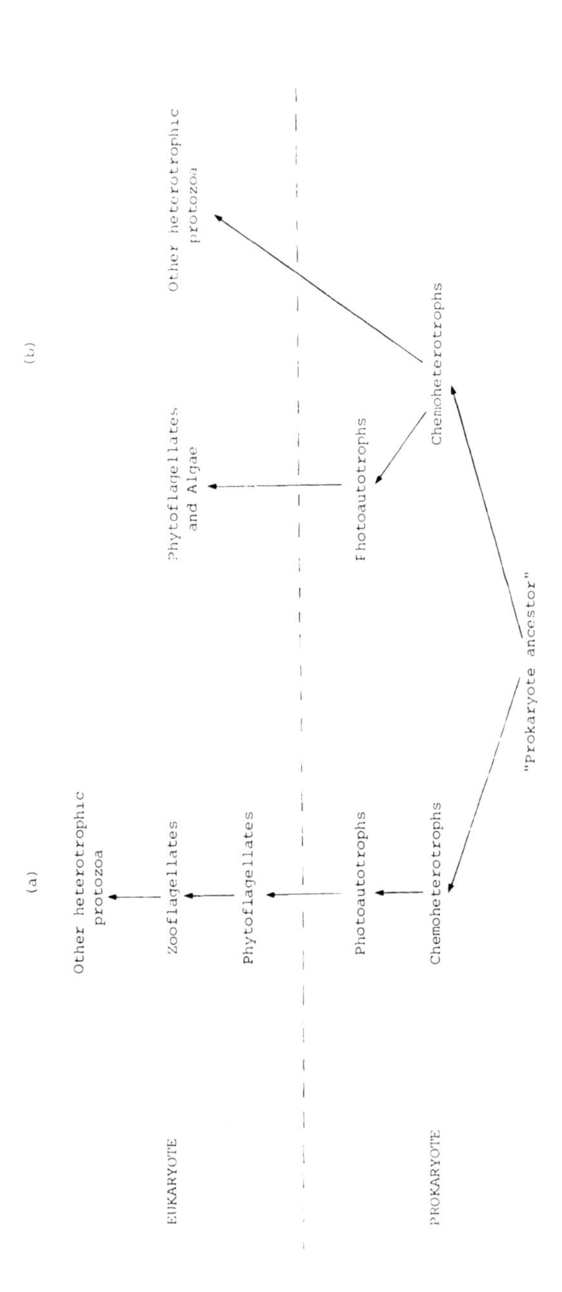

Figure 5.1: Two Possible Views of the Evolution of Nutritional Strategies. (a) The existence of mixed nutrition groups, e.g. phytoflagellate euglenids, is explained by loss of pigment in some to give transitional groups between phyto- and zooflagellates; (b) mixed nutrition groups are phytoflagellates but revealing heterotrophic ancestry.

Figure 5.2: Evolution of Nutritional Strategies by the Endosymbiotic Theory. * indicates where prokaryote endosymbionts may have been acquired, first to add eukaryote cell organelles, then photosynthetic symbionts as a more recent event.

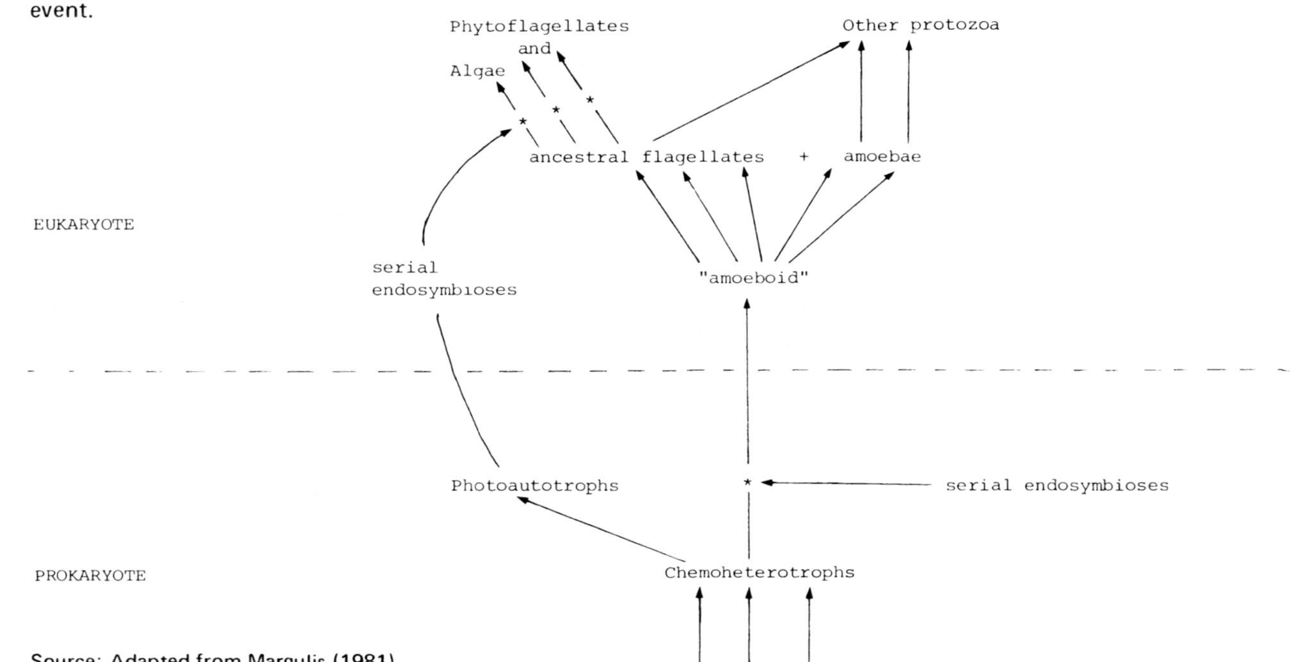

Source: Adapted from Margulis (1981).

and light-receptive pigments not found in the majority of protozoa. Chemoheterotrophic protozoa then arise directly from their prokaryote ancestors in this scheme (Figure 5.1b).

A view now becoming widely accepted is that eukaryotic cells evolved when simple prokaryotes acquired a series of other prokaryotes as endosymbionts. These endosymbionts become progressively more integrated into their host cells, establishing complete interdependence with them, and eventually becoming the membrane-bound organelles now recognized as constituents of eukaryote organisms. Taylor (1974) describes this theory of eukaryote evolution as the Serial Endosymbiotic Theory. Margulis (1981), for many years a supporter of this line of thinking, expands and uses it to present her view of the evolution of nutritional strategies in eukaryote cells (Figure 5.2). The prokaryote heterotroph engulfed endosymbionts, gradually evolving into an 'amoeboid' ancestor of the eukaryotes. By a further series of endosymbiotic events this amoeboid heterotroph acquired photosynthetic prokaryotes to become photoautotrophic flagellates and algae. The 'amoeboid' ancestor, already established as a chemoheterotroph, could have developed locomotory and other distinguishing organelles by further endosymbioses. There is evidence amongst extant forms of locomotion being mediated by ectosymbionts (Cleveland and Grimstone, 1964; Tamm, 1982).

Whatever the view of nutritional strategies adopted, one conclusion is common to all: that most protozoa are heterotrophs. This gives heterotrophic nutrition a prominent position in a general review of feeding methods. The following three chapters concern themselves mainly with heterotrophy. This chapter discusses some of the ways in which nutrients enter the organism; the next follows with how nutrients are contained within the organism and finally, some outline of the biochemical breakdown processes.

Endocytosis

Heterotrophic protozoa, in common with larger animals, require preformed compounds, either complex molecules or their simpler breakdown products. These food components may be particulate, semi-solid or in solution; thus the means by which food material enters the body is to a great extent dependent on the size of the particles involved. Some form of endocytosis is employed when the material ingested is enclosed in a sac formed from part of the plasma membrane of the organism. Therefore, the food does not pass directly into the cytoplasm but into a food vacuole whose membrane wall keeps it separate from the cell constituents. Thus it remains technically a part of the outside world, i.e. extracellular. It has been maintained by de Duve (1963) that digestion is intracellular, that the endocytotic (food) vacuole is equivalent to an intracellular digestive tract in physiological terms, and that it corresponds to the digestive tract of a metazoan: a debatable point.

The various forms of endocytosis (or food uptake) really only differ in degree. The process is essentially the same whether it is phagocytosis of particles or other organisms or pinocytosis of molecules. Accidental uptake of substances must also occur during both processes. Both endocytotic processes are affected identically by inhibitors of aerobic metabolism and by low temperatures. They are closely linked in that the combined volume taken up by endocytosis is constant and critical. In *Acanthamoeba castellani* this represents 15 per cent of the cell volume (Bowers, 1977). As phagocytosis increases, pinocytosis must decrease proportionately.

In addition to the forms of bulk transport just considered, which involve invagination of the plasma membrane, other essential substances, dissolved nutrients of low molecular weight, enter the organism by facilitated diffusion or active transport through the plasma membrane. Some protozoa secrete hydrolyzing enzymes into the external medium to degrade large nutritive molecules into smaller soluble units for transport through the plasma membrane. *Tetrahymena* produces lysosomes specially for export when maintained in a nutritive, liquid-only medium such as proteose peptone broth (Blum and Rothstein, 1975). This facility for extracellular digestion is of value to facultative and obligate parasites and to other protozoa which live in a highly nutritive environment.

Phagocytosis

In discussing phagocytosis as a means of acquiring food, certain factors must be considered in general terms at this point. They come under two broad questions: (a) how is the food collected and (b) what stimulates a protozoan to feed and perhaps select its food?

Uptake of nutrients by phagocytosis involves a wide diversity of mechanisms in protozoa, although a cytostome or fixed point of intake is not an essential requirement (Figure 1.2). Equally a wide diversity of food is taken up with bacteria, small algae and detritus being the most commonly acceptable. For this reason methods of extracting small particles from the medium are much in evidence particularly amongst ciliates.

The efficiency of extracting small particles by filtering quantities of water depends on the size of the spaces between adjacent cilia or ciliary membranelles. Species which are adapted for feeding on small particles, such as bacteria of size 0.2-1.0 μm, may not clear a suspension so efficiently because the small sieve spaces necessary to trap these bacteria also result in decreased water flow through the filter (Fenchel, 1980a, b, d).

Food Gathering in Ciliates. Elaborate organelles for filter-feeding have evolved, particularly amongst ciliated protozoa. Some of these will be described more fully in Chapter 8. The small, fast-moving hymenostome ciliates generally have ciliary membranelles on the left side of the oral cavity, beating metachronally and driving water dorsal and right against an undulating membrane (a haplokinety), which collects food particles for phagocytosis. Slightly different is the arrangement in the hymenostome *Glaucoma* which has three ciliary membranelles lying in its oral cavity. Parallel membranelles 1 and 2 beat metachronally but out of phase, creating temporary pools of 'dead' water between them, which is then squeezed posterior and right, through filtering membranelle 3 (Fenchel and Small, 1980, and Figure 5.3b). These membranelles can circulate 4000 μm^3 of water through the oral cavity of *Glaucoma* every second.

The slower browsing spirotrich ciliates use a band of perpendicular ciliary membranelles to drive water backwards to the posterior end of the cytostome. Particles of less than 1-2 μm will not be sieved out of the water current by the spirotrich *Euplotes moebiusi*, for the space between adjacent membranelles is about 1.5 μm (Fenchel, 1980b). The petritrich *Vorticella* collects food by the combined action of two bands of cilia: the double inner polykinety draws water and particles in by its propulsive action and the outer, less active haplokinety acts as a filter (Sleigh and Barlow, 1976, and Figure 5.3a).

The morphology of feeding organelles and the part they play in size selection during feeding and feeding rate, are all of importance in sorting ciliates into favourable ecological niches. Bactivorous ciliates cannot live in open water as they need a high concentration of small bacteria in order to trap enough for their requirements. Tintinnids and oligotrichs are important members of plankton for, as large particle feeders, they can graze on phytoplankton (Fenchel, 1980b, c).

Figure 5.3: Food Trapping by Ciliary and Flagellar Action. The direction of water currents initiated by cilia or flagella is shown by the dashed lines and arrows. (a) *Vorticella*, (b) *Glaucoma*, (c) *Actinomonas*, (d) *Codonosiga*.

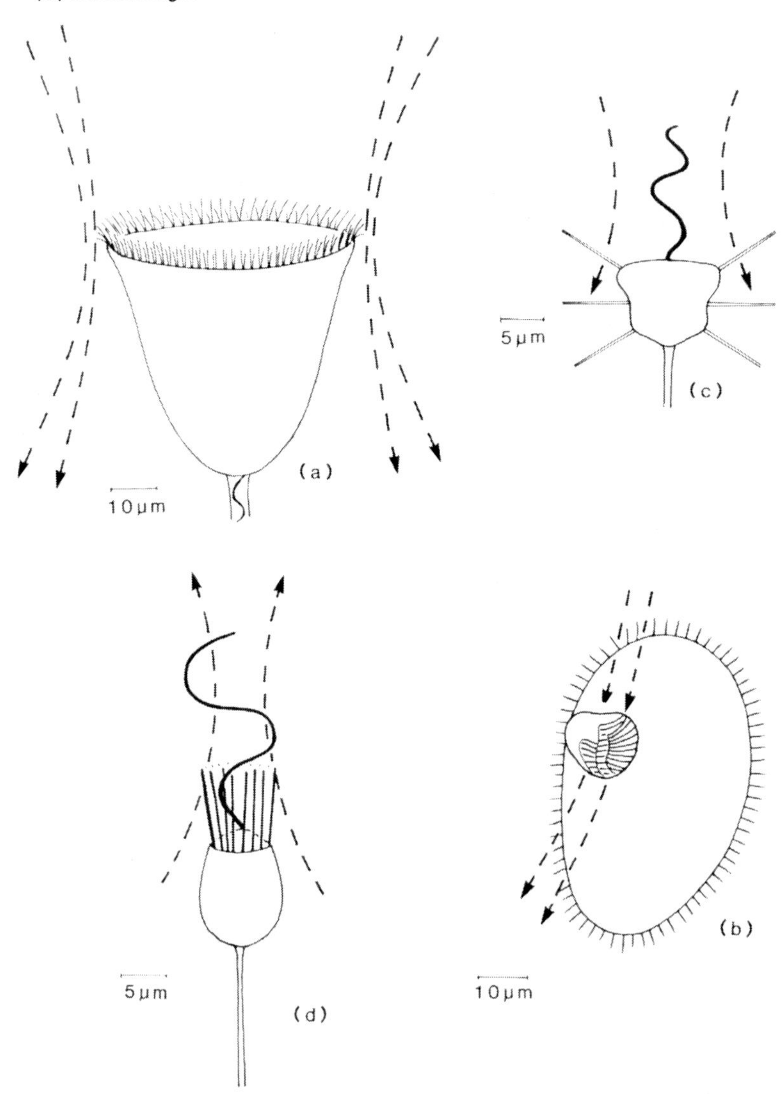

Source: Redrawn from: (a) Sleigh and Barlow (1976); (b) Fenchel and Small (1980); (c) and (d) Sleigh (1964).

Food Gathering in Flagellates. The use of flagella in trapping fine food particles is not so well documented. Sleigh (1964) described the way in which flagellar undulations create feeding currents in some small, semi-sessile flagellates. Although their flagella produce plane sinusoidal undulations passing from base to tip, amongst the species he considered only in *Codonosiga* do the water currents follow the same direction (Figure 5.3d). In *Actinomonas* and *Monas* water currents pass in the opposite direction, from tip to base of the flagellum (Figure 5.3c).

Figure 5.4: Mastigonemes on the Primary Flagellum of *Monas*. Hair-like projections act as oars to drive the swimming/feeding current backwards (x 10,000).

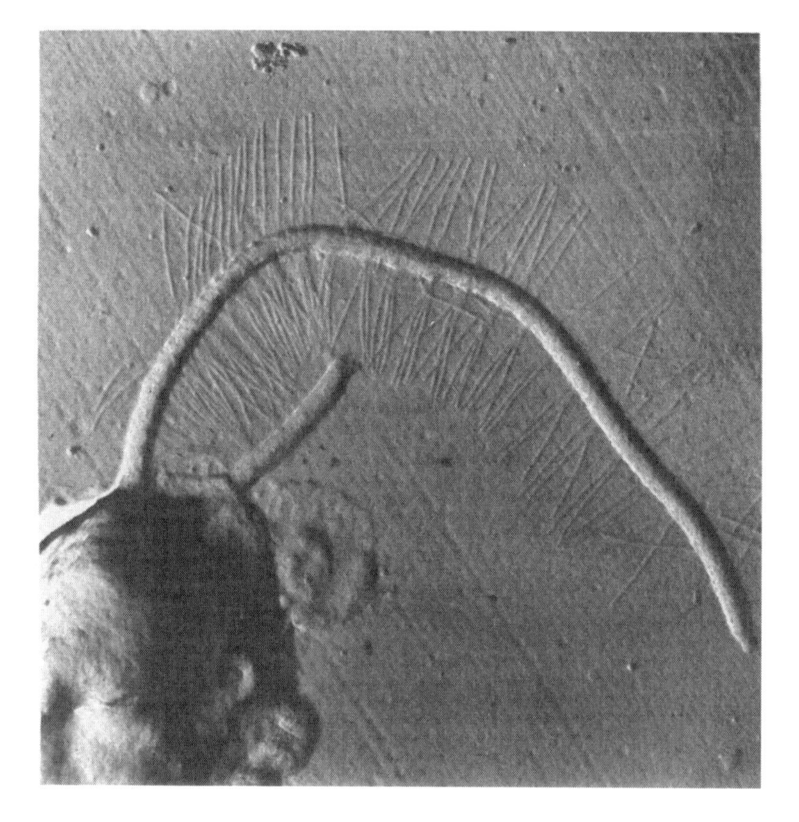

This backward flow of water may be caused by mastigonemes which project down both sides of the flagellum acting as oars 'rowing' water backwards when fully extended at the crest of each flagellar wave

(Figure 5.4). A similar principle operates in polychaete worms which extend their parapodia to provide a power stroke during undulatory swimming movements (Trueman, 1975).

Each flagellate creates water currents which conduct small particles to the point on its surface where phagocytosis can take place. In *Monas*, food particles must be funnelled in to contact the plasma membrane near the flagellar base; *Actinomonas* has food-catching pseudopodia (filopodia) and needs a broader water current. *Codonosiga* has a projecting collar, a palisade of microvilli, through which food particles pass as the water is drawn forwards over it. These particles drop into the base of the collar where they are ingested (Figure 5.3d).

Selection and Induction. Amongst the filter-feeding protozoa inert latex beads are accepted as readily as bacteria of comparable size. It seems here that size not quality is the basis for selection (Fenchel, 1980b). However, chemoreception undoubtedly plays a key role in enabling some protozoa to select suitable food, particularly amongst carnivores which engulf whole organisms.

Surface contact which results in capping is an important aid to feeding in some amoebae. Receptors on the plasma membrane of *Acanthamoeba* respond to the motile bacteria *Pseudomonas* by binding them in large numbers to its surface through their polar flagella (Preston *et al*., 1982). As *Acanthamoeba* only engulfs food at its posterior (uroid) end, bacteria must be moved towards this point. This action is achieved by aggregates of bacteria being moved as a cap posteriorly, followed by endocytosis. More bacteria cannot bind to an already capped *Acanthamoeba* as the flagellar binding sites have become temporarily depleted. Nor can they bind to an amoeba which has been pre-treated with a plant lectin such as Concanavalin A which blocks the carbohydrate binding sites on the plasma membrane glycoproteins (Preston *et al*., 1982).

The nature of binding sites on the protozoan plasma membrane is clearly of importance and interest. Sugars as specific binding sites in the glycoprotein layer of the protozoan cell membrane are being identified by many investigators. Galactose and mannose residues are active in *Euglena*, *Chlamydomonas* and *Tetrahymena* (Sharabi and Gilboa-Garber, 1980).

A wide range of inducer substances is implicated in feeding in different protozoa (Seravin and Orlovskaja, 1973, 1977). Some chemicals which induce phagocytosis in free-living amoebae include hyaluronidase, trypsin, cytochrome C, peptones and lecithin, and in the ciliate, *Coleps*, a variety of SH-bearing substances such as glutathione and cysteine, and phospholipids. The carnivorous flagellate, *Peranema*, is also stimulated to feed by phospholipids such as lecithin and cephaline and by trypsin and tweens.

Once food has been accepted, it is engulfed by some form of invagination or phagocytosis. There are grounds for confusion in terminology between phagocytosis and pinocytosis, particularly in parasites of cells and nutrient body fluids. The phagocytic vacuoles of *Entamoeba histolytica* can only be distinguished from macropinocytotic vacuoles at the ultrastructural level, because the former contain fragments of host cell debris (Martinez-Palomo, 1982).

Pinocytosis

Small molecules such as proteins are taken up in solution by pinocytosis and enter the organism in small vesicles. In an amoeba pinocytotic vesicles are formed at the bases of long narrow invaginations, pinocytotic channels. Small vesicles are pinched off at the base of a channel deep in the cytoplasm and are passed into the interior. But pinocytosis does not necessarily involve the development of channels. The parasitic *Opalina ranarum*, which must take up nutrients through its plasma membrane as it has no mouth, does not form channels; it pinches off small vesicles in the grooves between the folds in its pellicle (Münch, 1970). Once inside, vesicles coalesce to form larger vacuoles bounded by unit membranes, vacuoles which are not markedly different from endocytotic vacuoles produced by phagocytosis.

Pinocytosis may occur simultaneously all over the surface as in amoebae, or it may be restricted to clearly defined regions, such as the walls of the flagellar pocket of some trypanosomes. In protozoa which have a sculptured pellicle such as is normally found in ciliates, pinocytosis is confined to grooves, hollows or pits particularly round the bases of the cilia. However the well-defined parasomal sacs which exist as permanent invaginations of the surface membrane adjacent to the ciliary basal bodies in many ciliates, have not yet been clearly implicated in pinocytosis. One might question whether, if permanent structures such as parasomal sacs are used for pinocytosis, this is strictly pinocytosis.

Suctorian ciliates, predators which absorb bulk food through their suctorial tentacles, have the capacity to take in growth requirements through pits on the surface of the body and on the tentacular shafts (Rudzinska, 1980). The elaborate cortical structure in *Tokophrya* means that the pits are the only possible sites of pinocytosis in these suctorians. These pits are lined only by the outer pellicular membrane. The extraneous amorphous coat, the two inner pellicular membranes and the epiplasm are disrupted at these points, making an opening which leads through a narrow neck to a single-membrane lined saccule (Figure 5.5). Flattened pinocytotic vesicles are pinched off from the wall of the saccule and they show an unusual tendency (for pinocytotic vesicles) to stack in piles rather reminiscent of Golgi structures.

Figure 5.5: Permanent Pinocytotic Pits in *Tokophrya*. A diagrammatic section through the cortex of *Tokophrya* shows a pinocytotic vesicle separating from the saccule of the pinocytotic pit, which is lined by outer pellicle membrane only. ac, extraneous amorphous coat; ep, epiplasm layer of cortex; pm, three pellicle membranes of outer cortex; sa, saccule.

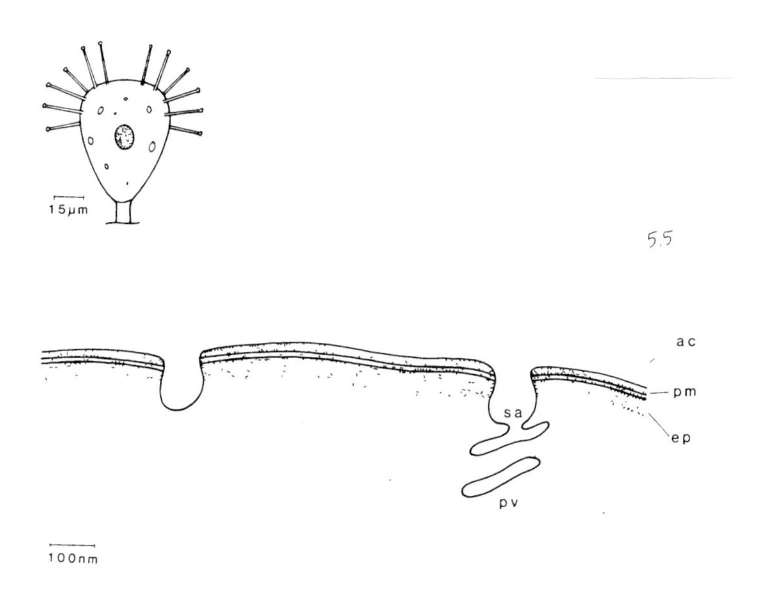

Source: Adapted from Rudzinska (1980).

Induction Mechanisms in Pinocytosis

The rather simplified view of pinocytosis just presented hides a complexity which is only now being elucidated by membrane biologists.

It has been known for many years that inducer substances are necessary for successful pinocytosis. It has been found experimentally that pinocytosis can be induced in amoebae by a large number of ionized compounds that react with constituents of the glycocalyx (mucopolysaccharide coat) overlying the plasmalemma (Chapman-Andresen, 1962). As a mode of entry of soluble molecules into cells pinocytosis is now recognized as a widespread phenomenon. However, Chapman-Andresen (1973) argues that the amount of solute taken up by pinocytosis varies according to the inducer.

Pinocytosis can be induced in *Amoeba proteus* by substances of lower molecular weight than in mammalian cells. Amino acids, sodium chloride and other inorganic salts, dilute sea water, basic dyes such as Acridine Orange and Alcian Blue, and proteins are recognized as inducers. As induction is a surface reaction, the criterion of a successful inducer is that the molecules must be positively charged and must bind to the surface. Then follows a sequence of easily observable events which are well-documented: cessation of movement, rosette formation, the development of invagination channels and pinocytotic vesicles.

The processes of induction and channel formation are not fully understood, but work by Prusch and his colleague (1979, 1980) on *Amoeba proteus* is beginning to throw some light on the process. Using 0.01 per cent Alcian Blue as an inducer, they found that the presence of calcium ions in the external medium was also essential and that a maximum rate of pinocytosis could be reached in a concentration of 10^{-4} M Ca^{2+}. The effect of the inducer, in this case Alcian Blue, is to displace bound calcium ions from the outer surface of the plasmalemma of *Amoeba proteus*, releasing them into the external medium. These free Ca^{2+} ions may now pass through the plasmalemma into the cytoplasm, by active transport (see below). The localization of Ca^{2+} below the plasmalemma is important in pinocytosis as it stimulates the contraction of microfibrils, contractile elements which are important in rosette formation, invagination channels and vesicle separation. Following the release of calcium ions Alcian Blue binds to the surface, an action which initiates the production of pinocytotic channels by surface invagination. Adsorbed Alcian Blue inducer and bulk medium with its constituent solute molecules such as sucrose are pinched off as pinocytotic vesicles which pass into the cytoplasm. As the plasmalemma is normally impermeable to sucrose and other carbohydrate molecules which cannot themselves induce pinocytosis, then the action of an inducer provides an important mode of entry. Recent experiments with the carrier antibiotic, calcium ionophore A 23187, have shown that the complex which it forms with Ca^{2+} can migrate directly across the plasmalemma of *Amoeba*, releasing Ca^{2+} inside and so short-circuiting the normal sequence of events (Figure 5.6). As the influx of sucrose into *Amoeba proteus* is dependent on the internal localized accumulation of charged calcium ions then any means such as the use of calcium ionophore, which speeds up the entry of calcium ions into the cell, will also enhance the rate of entry of sucrose (Prusch, 1980).

In common with other methods of feeding, pinocytosis is not a continuous activity. A cycle of pinocytosis depends on the physiological state of the organism and on the type of inducer. *Amoeba proteus*, forming pinocytotic channels all over its surface during pinocytosis, has an extensive utilization of plasmalemma; up to 70 per cent of this membrane may be involved in a 30-minute cycle. Measured in terms of high energy requirement and membrane turnover, active pinocytosis is of doubtful use to many protozoa, being much more costly than the engulfment of a single food organism such as the ciliate *Tetrahymena*.

Figure 5.6: The Involvement of Inducers and Calcium Ions in Pinocytosis. (a) Calcium ionophore – Ca^{2+} complex, with its hydrophobic and lipid soluble exterior, migrates directly through the plasma membrane releasing Ca^{2+} inside; (b) Alcian Blue displaces Ca^{2+} and binds to the plasma membrane; released Ca^{2+} enters the cell through calcium channels; (c) Ca^{2+} accumulates below the plasma membrane to facilitate the contractile processes necessary for invagination and separation of pinocytotic vesicles (pv).

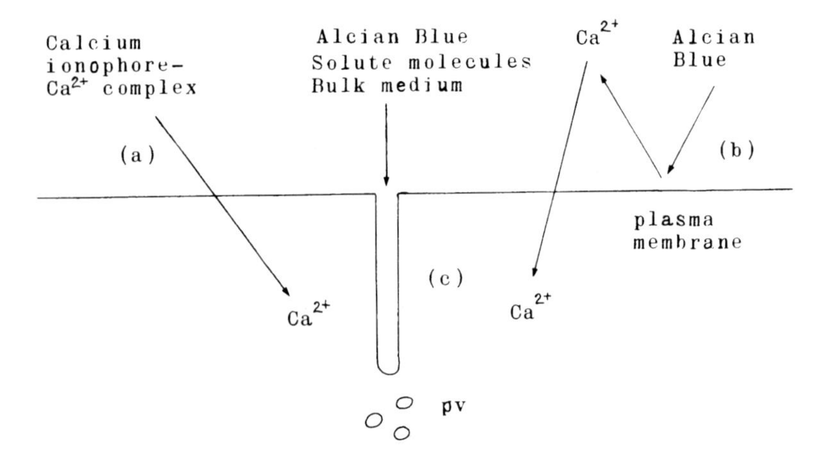

Membrane Transport of Small Molecules and Ions

Entry of individual small molecules into the organism by facilitated diffusion or active transport differs from phagocytosis and pinocytosis in that here molecules pass through the membrane structure itself. Although these modes of entry are characteristic of both eukaryote and prokaryote cells, they will be discussed here only in the context of the protozoan cell, drawing examples from current research using protozoa as model organisms.

Facilitated Diffusion

This involves the transport of selected solute molecules or ions into or out of the cell by specific carrier proteins which are an integral part of the membrane. Currently there are thought to be two ways in which transporter proteins could operate. The first is by forming permanent channels or pores through which molecules of a suitable size and charge can pass. In the alternative theory solute molecules bind to mobile carrier proteins which move across the membrane, picking up molecules on one side of the membrane and discharging them on the other. Whichever explanation provides the answer the result is the same – molecules pass in both directions under the influence of the concentration gradient and therefore with no expenditure of energy.

Active Transport

Active transport of molecules or ions, by contrast, is an energy-consuming process as it often operates against considerable concentration gradients. Much attention of protozoologists has been directed towards the transport of Ca^{2+} ions through the plasma membrane of *Paramecium* since it was shown that Ca^{2+} is responsible for initiating ciliary reversal or the avoiding reaction.

Paramecium has proved to be a very convenient model for the study of membrane transport as it is large enough for measurements to be made of internal and external concentrations of ions and of changes in surface charges (Browning and Nelson, 1976; Kusamran *et al.*, 1980). Work has also been carried out on *Didinium nasutum* (Hara and Asai, 1980), and *Tetrahymena* (Kusamran *et al.*, 1980). Browning and Nelson (1976) studied calcium transport through the ciliary membrane in *Paramecium aurelia*. Depolarization of the membrane occurs on physical contact or due to changes in the external ionic concentration (perhaps some noxious substance?). Calcium 'gates' open under the control of specific surface proteins, temporarily producing channels or pores, and Ca^{2+} rushes in because ciliates maintain an internal calcium level three orders of magnitude lower than the external medium. At the same time, K^+ ions diffuse out slowly as the internal potassium level is maintained slightly above the external concentration.

Figure 5.7: Transport of Ca^{2+} Ions through the Ciliary Membrane of *Paramecium*. (a) Calcium 'gates', controlled by specific surface protein molecules (ssp), open on depolarization in response to contact or external ionic change. Influx of ions is passive as the internal Ca^{2+} concentration is maintained at a lower level than the external medium; (b) Ca^{2+} is actively transported out against the concentration gradient, using ATP energy to operate the calcium pump.

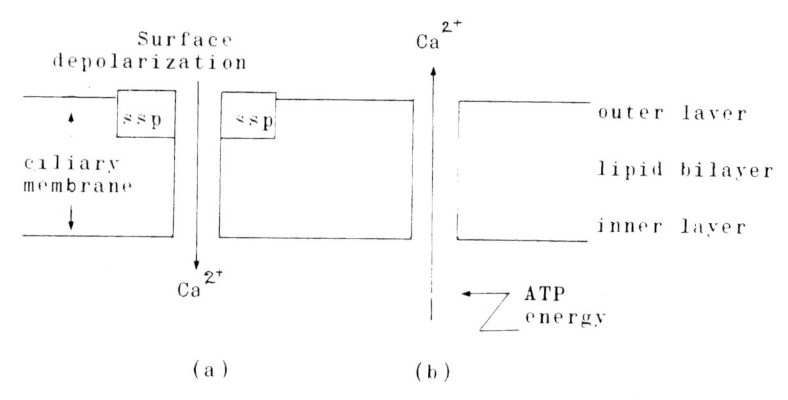

The ciliary reversal induced by the influx of Ca^{2+} is rapid and short-lived because the calcium pump comes into operation and actively transports Ca^{2+} out through the membrane again (Figure 5.7). Although the influx of Ca^{2+} into the organism is a passive process, its efflux requires energy as Ca^{2+} is pumped out against the concentration gradient. This energy is obtained from membrane-bound ATPase, the probable mechanism being that Ca^{2+} ions form a complex with the enzyme. It is also considered that the membrane lipid bilayer has an important part to play in active transport.

Ca^{2+} ions pass across the cell membrane through calcium channels, thereby inducing a change in ciliary activity in *Paramecium*. The precise location of these channels in relation to the cilia is clearly of interest. Voltage-sensitive calcium channels were shown to be localized on the surface membrane covering the cilia in *Paramecium caudatum* (Dunlap, 1977). By 'deciliating' an organism and monitoring the calcium response in relation to the regrowth of the cilia, the nil response of a deciliated

Paramecium did not revert to complete excitability until the cilia were back to their original length. Later work positioned the calcium channels more precisely at points just above the bases of the cilia (Kusamran *et al.*, 1980). Inevitably the behaviour of other related ions was investigated. Ling and Kung (1980) found that calcium channels could be opened in *Paramecium* on excitation with sodium or barium ions: Ba^{2+} enters the cells via the calcium channels, as barium is an analogue. However, the mechanism of active transport efflux is more specific and does not operate in this situation, with the result that Ba^{2+} was found to accumulate in the cells. Evidence for the involvement of calcium channels in ciliary reversal also comes from a mutant of *Paramecium aurelia*, the pawn mutant, which can only swim forwards as it has defective calcium channels.

Active transport is an important mechanism for the uptake of soluble metabolites as well as for inorganic ions. Mutant strains of *Tetrahymena pyriformis* which lose the capacity to form food vacuoles at 37°C continue to grow in a medium containing 2 per cent proteose peptone, high concentrations of vitamins and salts of heavy metals. There must be adequate modes of entry of nutrients in the absence of food vacuoles. Carrier-mediated active transport for the uptake of essential nutrients has been demonstrated in *Tetrahymena*. Considering the structural complexity of the pellicle layer in *Tetrahymena*, in common with many ciliates, it is not surprising to find that the membrane lining the cytopharyngeal pouch is the principal site of active transport (Dunham and Kropp, 1973). Uridine, leucine, alanine, amino acids, arabinose and glucose, with which it competes for a carrier, are all known to enter by this route. Corroboration that energy is required for carrier-mediated active transport in *Tetrahymena*, as with other organisms comes from the use of the metabolic inhibitor 2,4-dinitrophenol. The rate of entry of leucine is inhibited by 86-90 per cent and of alanine by 75 per cent in concentrations of 5×10^{-5} M 2,4-dinitrophenol.

Mediated active transport in parasitic protozoa, many of which live bathed in nutrient fluids of the host's body, and have no cytostomal opening, is of particular importance. Glucose and leucine enter *Trypanosoma lewisi* by active transport whilst K^+ ions enter by diffusion (Schraw and Vaughan, 1979). In trypanosomes such as *T. brucei*, which possesses a surface coat in the bloodstream phase of its life history, the coat has been shown to reduce the parasite's capacity to absorb fatty acids (Voorheis, 1980). *In vitro* culture forms of *T. brucei*, which have no surface coat as they require no protection from host antibodies, take up the fatty acids, oleic and stearic, demonstrably more efficiently than bloodstream forms. As uptake of fatty acids is a two-stage process, rapid initial binding to the plasma membrane followed by a slower accumulation within the organism, it would seem that the surface coat masks some of the possible binding sites, so making it more difficult for coated trypanosomes to absorb fatty acids.

From this brief consideration of ways in which nutrients enter heterotrophic organisms, one fact emerges – that the plasma membrane has an important role to play. Whether this role involves the actual removal of sections of membrane from the outer surface as vacuoles and vesicles migrate to the cytoplasmic interior or the *in situ* activity of the membrane during facilitated diffusion and active transport depends on the size and molecular configuration of the nutrient material. Invagination is expensive in terms of membrane utilization. The organism cannot pinch off endocytotic vacuoles faster than it can produce replacement plasma membrane. A starved protozoon when presented with food does not necessarily launch into a frenzy of feeding activity. Some protozoa take time to build up reserves of energy and membrane constituents; others do have a reserve of specialized membrane with which to form endocytotic vacuoles. Even in a normally feeding protozoon continual re-synthesis of membrane units cannot always proceed fast enough to keep pace with requirements. How the organism solves this problem of membrane recruitment will be discussed more fully in the next chapter.

In Summary

It was recognized at the beginning of this chapter that heterotrophic nutrition in protozoa is dependent on a steady supply of complex organic molecules. Whether this food material enters the organism by phagocytosis or by pinocytosis, it becomes contained in endocytotic vacuoles in which digestion proceeds. Methods of gathering particulate material in ciliates and flagellates, and the importance of inducers of pinocytosis, have been discussed. Other transport mechanisms, facilitated diffusion and active transport, although not restricted to protozoa, have been discussed in the context of protozoan examples.

Variation in the modes of formation of vacuoles and the process of digestion will be discussed in the following chapters.

6 ENDOCYTOTIC VACUOLES IN DIGESTION

Heterotrophic protozoa collect or capture their food in a variety of ways, involving versions of phagocytosis, pinocytosis and membrane-mediated events. This aspect has been considered in some detail in the previous chapter. By whatever means food material enters the organism, much of it will be located initially within an endocytotic vacuole, alternatively referred to as a food vacuole or a secondary lysosome. In fact the whole process of digestion is centred round the endocytotic vacuole, which is produced by being pinched off from the base of the cytopharynx, the vestibule, the suctorial tentacle or other site used for food intake.

Figure 6.1: Ciliate *Pseudomicrothorax* Ingesting a Filament of *Oscillatoria*. Scanning electron micrograph by R. Peck (x 1,300).

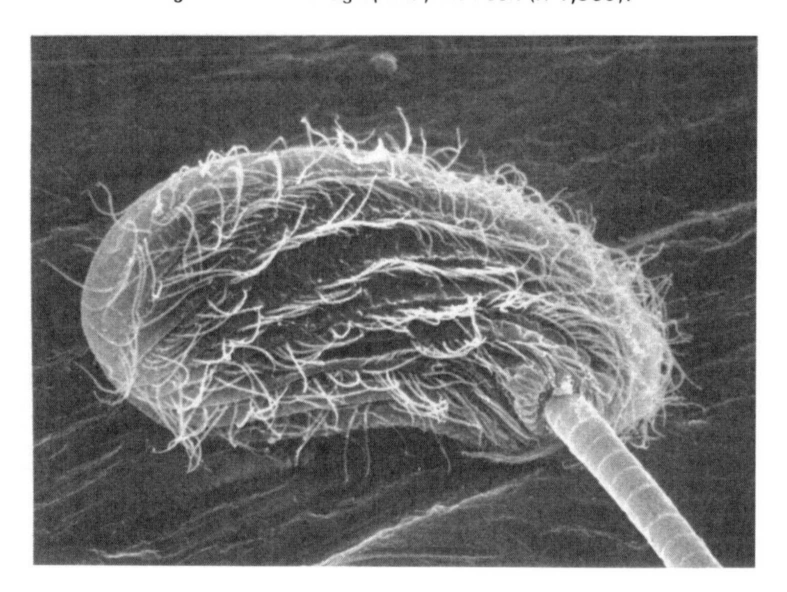

The production of endocytotic vacuoles during feeding requires a supply of membrane to allow for vacuole enlargement. Newly forming vacuoles may be large and not necessarily spherical at first if the food taken in is bulky or irregular in shape. Some herbivorous ciliates take long sections of filamentous cyanobacteria (Figure 6.1). *Pseudomicrothorax* ingests filaments of *Oscillatoria* (Hausmann and Peck, 1979; Peck and Hausmann, 1980) and *Nassula* takes filamentous *Phormidium* (Tucker, 1968), where bulky food must be enclosed until partial lysis of the filament makes the shape more manageable.

A problem of a different kind faces predatory protozoa which ingest particularly active food. The sessile spirotrich ciliate *Stentor* ingests active rotifers, *Monostyla*. By vigorous movements the rotifer may break out of the endocytotic vacuole and need re-enclosing in a new vacuole if digestion is to proceed, or it can escape completely through the outer pellicle membrane (Grula and Bovee, 1977). Although this is a rather extreme and unusual example of over-active prey, it is a common occurrence for small prey organisms to remain active for a time after ingestion.

Whatever the shape or size of the endocytotic vacuole, there follows a sequence of events starting with the collection of food items and culminating in defaecation of unwanted inert or indigestible remains. At some stage the vacuole receives a full set of enzymes capable of breaking down carbohydrates, proteins, lipids and nucleic acids. The presence of enzymes can be demonstrated by standard histochemical reactions (Gomori, 1952). Acid phosphatase used as an indicator of the presence of other lytic enzymes has been confirmed in endocytotic vacuoles of *Tetrahymena* (Elliott and Clemmons, 1966), *Paramecium* (Jurand, 1961; Meier *et al.*, 1980), *Tokophrya* (Rudzinska, 1972) and *Pseudomicrothorax* (Peck and Hausmann, 1980), to mention only a few.

The Digestive Process

Because of its dominant role in the digestive process the endocytotic vacuole must inevitably be a highly dynamic organelle capable of circulating by cyclosis and changing size by exocytosis and endocytosis as digestion proceeds. There have been many detailed studies of the formation and behaviour of vacuoles formed during feeding, following size changes, pH changes, the secretion of hydrolytic enzymes and the site of discharge of the contents of the residual vacuole. Descriptions of endocytotic vacuole events in *Tetrahymena* may be found in Elliott and Clemmons (1966) and Nilsson (1977), in *Paramecium* (Allen and Staehelin, 1981; Fok *et al.*, 1982; Jurand, 1961), in *Climacostomum* (Fischer-Defoy and Hausmann, 1977), in *Tokophrya* (Rudzinska, 1970,

1972), in peritrichs (Allen, 1978) and in amoebae (Roth, 1960; Stockem, 1973). In addition to the specific studies just listed, endocytosis in *Tetrahymena* has recently been reviewed by Nilsson (1979), as well as in opalinids (Wessenberg, 1978), in *Entamoeba histolytica* (Martinez-Palomo, 1982) and in amoebae (Chapman-Andresen, 1973).

Formation of Endocytotic Vacuoles

The details vary from species to species, but the basic process can be summarized diagrammatically using the peritrich ciliate, *Carchesium polypinum*, as an example (Figure 6.2). Food (bacteria and particles of detritus) is collected in the buccal cavity and directed down the cytopharynx to the cytostome. This arrangement and functioning of food-trapping organelles has been described by Sleigh and Barlow (1976) in *Vorticella*, but observations on the related *Carchesium* suggest that a similar pattern exists in it also. At the cytostome an endocytotic vacuole forms, enlarging as food accumulates within it and finally being pinched off by constriction.

Rate of Formation. Endocytotic vacuoles are formed at rates which vary with the size of the particles presented and the physiological state and age of the organism. Particle feeders such as *Paramecium* and *Tetrahymena* take up very small polystyrene latex beads of 0.3 μm diameter more rapidly than yeast cells of 4 μm diameter (Fok *et al.*, 1982; Nilsson, 1977). Size is apparently the controlling factor, rather than the composition of the particles, for *Paramecium* will take up bacteria or bacteria-sized latex particles at an equal rate (Müller *et al.*, 1965). The effect of age is reflected in experiments on *Paramecium* where organisms in the exponential growth phase will ingest food particles more rapidly than either younger or older organisms, for they are generally in a more active state of metabolism (Fok *et al.*, 1981, 1982).

Rates of feeding may be surprisingly high. The heterotrich ciliate *Climacostomum* can promote a very high feeding rate when presented with the small flagellate *Chlorogonium* or with yeast cells. *Climacostomum* is capable of ingesting 140 flagellates in 2.5 minutes, first by forming one very large vacuole, followed by a succession of smaller ones (Fischer-Defoy and Hausmann, 1977).

Similar rapid bursts of feeding can be induced in *Epistylis* which result in 20 food vacuoles being produced in 3 minutes (McKanna, 1973a). A more normal rate for steady vacuole formation in peritrichs such as *Carchesium* is one vacuole every 47-50 seconds, which con-

Figure 6.2: Cyclic Changes in Endocytotic Vacuoles in *Carchesium*. Digestive vacuole I (DVI): (a) elliptical nascent vacuole during its initial propulsive movement, (b) rounds up releasing surplus membrane as cup-shaped coated vesicles for return to the cytostome/cytopharynx. DVII: (c) condensing and shrinking due to water loss, (d) fully condensed vacuole receives enzyme-containing primary lysosomes. DVIII: (e) expanding vacuole assumes a wavy outline as its contents are digested, (f) mature vacuole pinches off secondary vacuoles with the products of digestion. Egestion vacuole (EV) discharges waste at the buccal cavity (g) bc, buccal cavity; cp, cytopharynx; csv, cup-shaped coated vesicle; cy, cytostome; hk, haplokinety; pk, polykinety; pl, primary lysosome; sv, secondary vacuole.

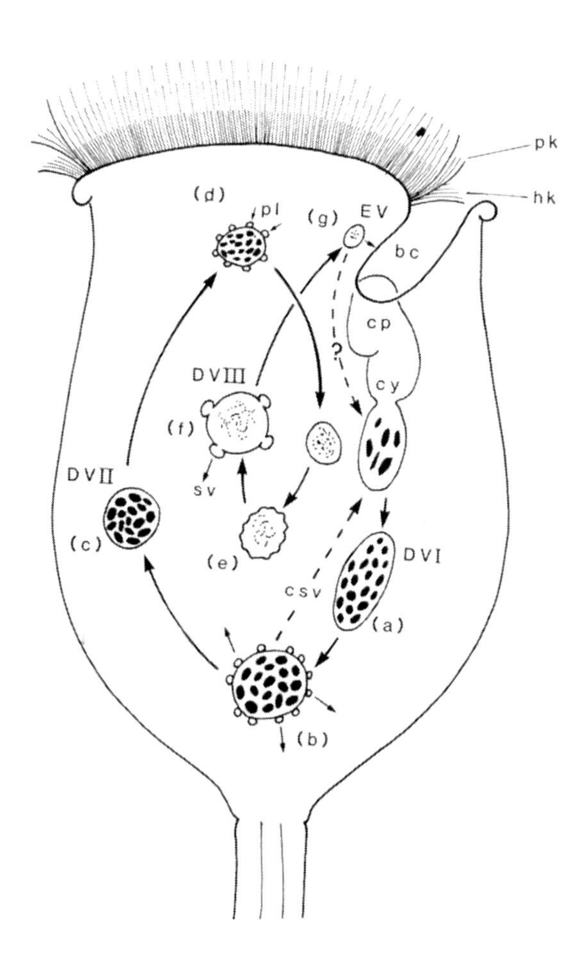

tinues whether the medium contains particulate material or not (Table 6.1).

Table 6.1: Rate of Formation of Endocytotic Vacuoles in *Carchesium polypinum* during Continuous Irrigation by Different Media.

Medium	*Average interval (secs)	S.D.
Tap water	49.8	± 2.04
Final sewage effluent	47.9	± 1.70
50 ppm Indian Ink in effluent	49.2	± 1.75

* Average time interval was calculated from sets of ten consecutive vacuoles, the time being recorded at the moment of release.

Source: Sugden (1950).

Phagocytosis by pseudopodial flow is necessarily a slower process than the ciliate method. Even so, *Chaos carolinense* may ingest 100 *Paramecium aurelia* in 24 hours, under optimum conditions (Christiansen and Marshall, 1965).

Separation of endocytotic vacuoles. The stimulus for the separation of the endocytotic vacuole is not fully understood, for it does not relate directly to the quantity of food in the vacuole (Table 6.1). There may be some effect due to the type or quality of the contents or to the extent of starvation of the organism. In many instances and in many organisms, digestive vacuoles are formed with wholly fluid contents; particulate matter is not essential. Given the appropriate stimulus the endocytotic vacuole is released from the ciliate cytostome by a constriction which pinches it from the membrane lining the oral region. This act must involve contractile elements (Nilsson, 1977).

The internal level of Ca^{2+} ions is thought to have an important role to play in the separation of the growing endocytotic vacuole, by initiating the contraction of contractile proteins. The involvement of ionic calcium in many physiological processes has been established beyond doubt, but its effect on phagocytosis has only recently been investigated. The macrostomal form of the ciliate *Tetrahymena vorax*, produced by heat shock from the microstomal form, captures *Tetrahymena pyriformis* in its cytopharyngeal pouch. Separation of the cytopharyngeal pouch to form a closed endocytotic vacuole may result from mechanical stimulation by active prey, causing depolarization of the pouch membrane and an influx of Ca^{2+} ions into the organism. Experimentally, devices for increasing the level of internal Ca^{2+}, such as the use of the cation ionophore A23187 which acts as a mobile carrier of Ca^{2+} across membranes, induce *Tetrahymena vorax* to form large empty vacuoles. 40 μM ionophore induces vacuole formation in 90 per cent of the treated organisms (Sherman *et al.*, 1982).

No precise figures are available for the level of internal Ca^{2+} ions required to initiate vacuole separation, but the site of the calcium-regulated contractile system has been postulated. This is thought to be the area of fine filamentous reticulum which lies below the right (ribbed) wall of the oral cavity in *Tetrahymena* (Williams and Bakowska, 1982, and Figure 8.3). Some of the interesting information recently published on structures associated with the oral region in ciliates will be discussed in Chapter 8.

Cyclic Changes during Digestion

Events involving the rapid growth of endocytotic vacuoles and the forces which direct their sudden movements indicate that the area of cytoplasm around the cytopharynx has special characteristics. These include (a) the concentration of disc-shaped vesicles whose function in vacuole growth will be discussed later, and (b) the microtubule bundles and post-oral fibre bundles which extend from the ribs of the cytopharyngeal wall, the lip of the cytostome and the undulating membrane, into the surrounding cytoplasm and perform various conducting functions (Allen, 1974; Williams and Bakowska, 1982).

Endocytotic vacuoles which are produced by the feeding organism go through a series of cyclic changes before the undigested residue is defaecated. To distinguish between the different stages in the digestive cycle it is convenient to name the endocytotic vacuole more precisely. The accepted nomenclature recognizes three stages of digestive vacuole, DVI, DVII and DVIII, and an egestion vacuole (Allen and Staehelin, 1981; Elliott and Clemmons, 1966; Fok *et al.*, 1982).

Digestive Vacuole I. This first stage includes the nascent, newly formed, elliptical or fusiform vacuole which separates from the base of the cytopharynx and glides quickly towards the posterior end of the organism (Figure 6.2(a)). Release of the newly formed digestive vacuole from the base of the cytopharynx and its sudden directional movement backwards is important because the way must be cleared for another vacuole to form in its place. Post-oral microtubule bundles or fibrils are clearly implicated in conducting the fusiform vacuole. On coming to rest temporarily the fusiform vacuole assumes a spherical shape (Figure 6.2(b)). In *Carchesium* only 15-20 seconds may have elapsed since the vacuole first started to form at the cytopharynx; *Paramecium* requires up to three minutes for this part of the digestive cycle.

The contents of DVI condense and surplus vacuole membrane is pinched off as cup-shaped, coated vesicles which migrate back to the cytopharynx; once there they transform into discoidal vesicles which are known to supply membrane for newly-growing digestive vacuoles (Figure 6.2(b)). This form of membrane transport in the ingestive-digestive cycle in ciliates has been described in *Paramecium* (Allen, 1974; Allen and Staehelin, 1981), in peritrichs (McKanna, 1973a, b), in *Tetrahymena* (Sattler and Staehelin, 1979) and in *Tokophrya* (Rudzinska, 1980).

As the digestive vacuole condenses, excess fluid is also removed by endocytosis of the cup-shaped vesicles, leaving a concentration of still metabolically-active bacteria or other food organisms. The contents of the digestive vacuole become progressively more acidic. Changes in pH in vacuoles vary from species to species; in *Paramecium* the pH may fall to 3.0. This increase in acidity may be fortuitous due to the respiratory activity of the ingested bacterial food, or it may be an active secretion designed to kill the food organisms and make the vacuole membrane more receptive to fusion by lysosomes (Fok *et al.*, 1982).

Digestive Vacuole II. The condensing vacuole gradually shrinks, perhaps to one-quarter of its original size and reaching maximum acidity (Figure 6.2(c), (d)). The time taken for this stage may vary considerably. Allen and Staehelin (1981) recorded 3-25 mins for *Paramecium* and Fok *et al.* (1982) found maximum shrinking and acidity to be achieved in *Paramecium* fed on polystyrene latex beads in 4-10 mins.

The newly-formed digestive vacuole contains no hydrolytic enzymes except perhaps traces from the ingested food. Generally it seems that the separation and the initial propulsive movement of the vacuole away from the base of the cytopharynx is the stimulus for release of enzyme-containing lysosomes from the endoplasmic reticulum, which will occur regardless of whether the ingested material is nutritive or inert polystyrene latex particles (Müller *et al.*, 1965). However, this generalization cannot be applied to all ciliates, for it seems that ciliates which ingest long filamentous cyanobacteria or algae behave differently. In *Pseudomicrothorax dubius*, feeding on filaments of *Oscillatoria*,

Hausmann and Peck (1979) have demonstrated that acid-phosphatase-containing primary lysosomes are added to the digestive vacuole as it forms, their function also being to contribute membrane to the enlarging vacuole. The problems facing *P. dubius* are to produce enough membrane in the cytopharyngeal region to enclose a long filament of *Oscillatoria* and to start the digestive process early so that the long filament may be accommodated quickly within the ciliate.

In the small particle feeders such as *Paramecium* and *Carchesium*, enzyme-containing primary lysosomes fuse with the fully condensed DVII (Allen and Staehelin, 1981; Fok *et al.*, 1982, and Figure 6.2(d)).

Digestive Vacuole III. As primary lysosomes fuse and digestion proceeds, the vacuole enlarges again, becoming less acidic or even slightly alkaline, and individual bacteria are no longer distinguishable. Expanding DVIII has an outline which now varies from wavy to irregular (Figure 6.2(e)). It has been reported that the time of maximum acid phosphatase activity does not coincide with the time of maximum acidity, but with the return to slight alkalinity (Müller, *et al.*, 1963). This rise in the intravacuolar pH of DVIII may be due to primary lysosomes performing a neutralizing function in addition to breaking down food material.

As DVIII matures it passes into a semi-quiescent period lasting 10-60 minutes or more when it moves little, and may shrink again through endocytosis of its fluid contents. The final breakdown of food is achieved through the agency of alkaline phosphatases, when the breakdown products are pinched off as small vesicles to be transported where necessary (Figure 6.2(f)). These small vesicles, or secondary vacuoles, have been observed in many protozoa in the later stages of digestion, and their presence may be used as a means of deciding the age of a digestive vacuole (Figure 6.3, and Jurand, 1961). Another means of deciding the age of digestive vacuoles has been elegantly demonstrated by Allen and Staehelin (1981) using freeze-fracture techniques in *Paramecium*. They showed that the structure of the vacuole membrane differed between DVI, II and III, particularly in the distribution of intramembrane particles.

The sequential size changes in the life of a digestive vacuole are shown in Figure 6.4. This emphasizes particularly the great reduction in diameter of DVII from the young DVI, then the gradual enlargement as primary lysosomes coalesce with the condensed vacuole.

Egestion Vacuole. Separation of small vesicles from the mature DVIII

Figure 6.3: Digestive Vacuole of *Paramecium aurelia*, Containing Partly Digested Bacteria (Ghosts) and Undigested Membranes, with Secondary Vacuoles in a Ring Outside the Vacuole Membrane. bg, bacteria ghosts; sv, secondary vacuole; vm, digestive vacuole membrane (x 7,500).

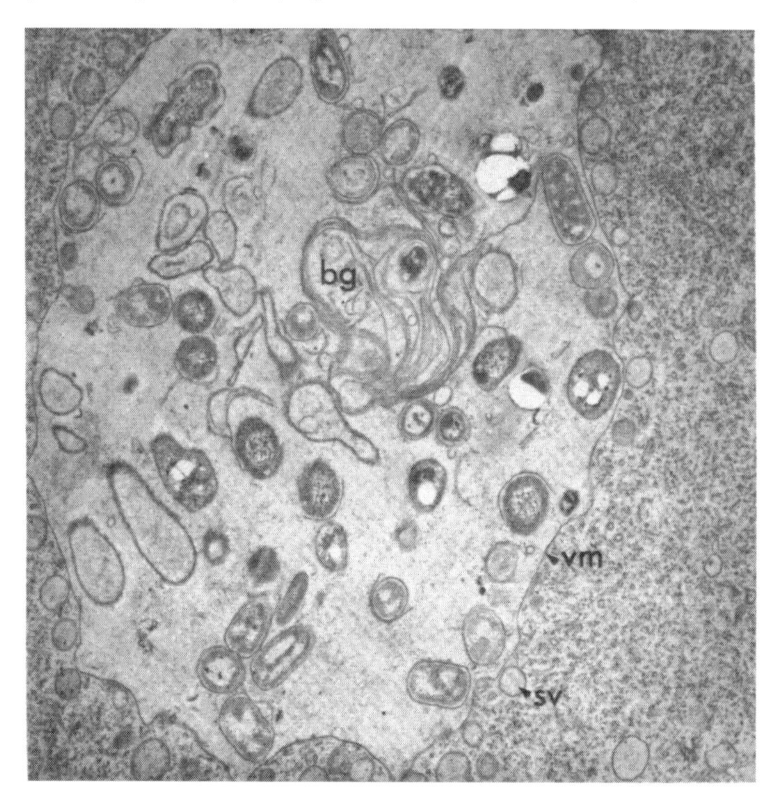

results in a reduction in size of the vacuole and a trend towards a neutral pH. Cyclotic movement of the digestive vacuole reaches its final stage. Any undigested remains, bacterial ghost membranes and other refractory materials, pass to the point of discharge in the egestion vacuole where they are defaecated through the cytopyge (Figure 6.2(g)). Membrane from the now spent egestion vacuole is thought to return in the form of small vesicles to the cytopharyngeal area for recycling (Allen, 1974; Allen and Staehelin, 1981). The position of the cytopyge in *Carchesium* is at a point high up in the buccal cavity, from which the remains are thrown off by reversal of tracts of the buccal ciliature; in *Paramecium* the cytopyge lies posterior to the oral groove. No generalizations can be made about the position of the cytopyge in ciliates except tentatively, that ciliates with an anterior mouth tend to have the cytopyge located at the posterior end.

Figure 6.4: Size Changes in the Life of a Digestive Vacuole. A large drop in vacuole diameter follows condensation and water loss between DVI and DVII.

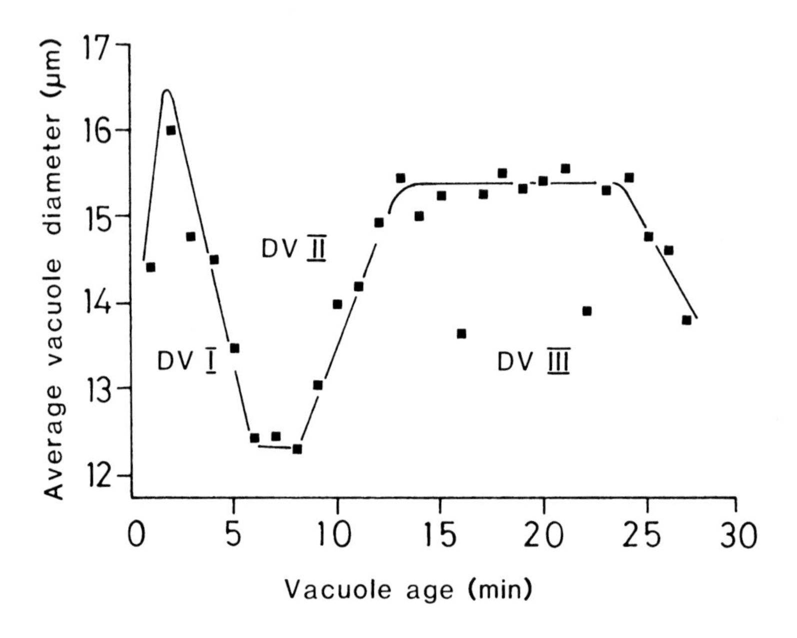

Source: Fok *et al.* (1982), by permission.

The time scale for the complete digestive process extends over a wide range; the length of time for the retention of inert particles is equally variable, as the following data show. *Paramecium* may void latex beads after 15 minutes, but others will be retained for an hour or more (Allen and Staehelin, 1981). *Climacostomum* passes latex beads and carmine particles through very quickly; yeast is mostly defaecated within 30 minutes but Indian ink particles can still be seen in digestive vacuoles 24 hours later (Fischer-Defoy and Hausmann, 1977). *Carchesium* egests Indian ink particles within 30 minutes (Sugden, 1950).

When nutritive material is presented to an organism, digestion and incorporation of its products may take place at an extremely rapid rate. *Euplotes,* when fed with *Tetrahymena* labelled with H^3-thymidine, is capable of transferring the labelled nucleotide to its own macronucleus in 2-4 minutes (Kimball and Prescott, 1962). The rates of production of digestive vacuoles in other ciliates have already been mentioned. When *Pseudomicrothorax dubius* ingests a cyanobacteria filament it wastes no time; within 5 seconds of entering the cytopharyngeal basket the end of the filament is lysed and becomes amorphous (Peck and Hausmann, 1980).

Mucus (mucopolysaccharide) is thought to play an important trapping role in the feeding process, especially in detritus- and bacteria-feeders (Curds, 1963; Sugden and Lloyd, 1950; Watson, 1945), although Fenchel (1980d) does not believe that mucus is involved when suspension-feeding ciliates trap their food. The ability to discharge mucus for feeding and movement is generally an important facet of invertebrate development. It occurs in many other invertebrate phyla, and in primitive chordates which employ ciliary currents for filtering small particles. Mucocysts, or mucus bodies, lie between the bases of the cilia in many ciliates and below the pellicle in some flagellates. In attached colonial peritrich ciliates mucus is an important aid in rejection currents. Unwanted particles are entangled by mucus into larger masses which can fall away from the vicinity of the colony more easily. The combined effects of ciliary currents and production of mucopolysaccharide have been illustrated in Figure 3.9.

It becomes clear from the dynamic events described during feeding that a great deal of membrane recycling must accompany it. As the major site of vacuole growth is the cytostome-cytopharynx region, this is where the recycled membrane units will accumulate, in the phagoplasm.

The Phagoplasm and Membrane Recycling

Further evaluation of the importance of the phagoplasm around the cytostome and of the various types of vesicles it contains, is relevant at this point. The need for quantities of replacement vacuole membrane during rapid feeding has been recognized. *Climacostomum* requires more than 1000 $\mu m^2/s$ when taking large prey (Fischer-Defoy and Hausmann, 1977). The impossibility of resynthesizing a sufficient quantity of membrane for phagocytosis (and rapid pinocytosis) highlights the importance of recycling the existing membrane, a process which is made easier because vacuole membrane retains its integrity and does not break up into molecular sub-units (McKanna, 1973a).

The principal sources of membrane are surplus components from condensing DVI and from the collapsed egestion vacuole (Figure 6.2). DVI represents a considerable source of membrane, for, on condensation, it may reduce its surface area by 39-50 per cent (Fok *et al.*, 1982; McKanna, 1973b). Surplus membrane separates from DVI as cup-shaped, coated vesicles which transform into discoidal vesicles in preparation for transport back to the cytostome-cytopharynx (Allen, 1974; McKanna, 1973a, b; Nilsson, 1977; Hausmann and Patterson, 1982).

Elaborate systems exist for returning discoidal vesicles to the cytopharyngeal region and assembling them for incorporation into the wall of a growing endocytotic vacuole. The pattern of assembly varies according to the shape of the food; a herbivorous ciliate which ingests long bacterial filaments produces a different shape of vacuole from a small particle feeder such as *Paramecium* (Peck and Hausmann, 1980). In *Paramecium* a group of microtubule ribbons, in bundles of 10-12 tubules attached to the left lip of the cytostome, act as a transport system (Allen, 1974). These microtubules filter discoidal vesicles from the cytoplasm and return them to the left lip where the new vacuole enlarges. Although Sattler and Staehelin (1979) confirm that disc-shaped vesicles coalesce with the growing food vacuole in *Tetrahymena*, they have been unable to identify a specific microtubule transport system for recycling membrane.

The suctorian ciliate, *Tokophrya,* which ingests food through suctorial tentacles located at various points over the surface of the organism shows no evidence of vesicle assembly or membrane recycling according to Rudzinska (1980). The implication is, though not proven, that the need for quantities of replacement membrane in localized areas is confined to the cytostome-bearing ciliates.

Recirculation of membrane components must be assumed to occur in amoebae (Bowers and Olszewski, 1972). An actively-phagocytosing *Acanthamoeba* can invaginate the equivalent of 2-6 times its surface area plasma membrane in a space of 30 minutes. Internalization of membrane by continuous pinocytosis results in a similar rate of turnover. Clearly the whole concept of membrane recycling is an interesting one and awaits more precise information from a wider range of protozoa.

Other types of vesicles in the phagoplasm are primary lysosomes carrying principally acid phosphatases, although several enzymes have been identified biochemically. In *Tetrahymena*, the enzyme content of primary lysosomes has been identified as belonging to two groups; those containing acid phosphatase, α-glucosidase, deoxyribonuclease, α-amylase and acetyl-glucosaminidase and others containing ribonuclease and proteinases (Müller, 1970; Müller *et al.*, 1966). There is evidence also that *Tetrahymena* receives pinocytotic vesicles which fuse with the digestive vacuole in the cytopharyngeal region (Elliott and Clemmons, 1966; Hill, 1972).

In Summary

Although this chapter has concentrated mainly on the formation and behaviour of endocytotic vacuoles as they occur in ciliates, other groups of heterotrophic protozoa, particularly the Sarcodina, show similar patterns of cyclosis.

The rates of formation of vacuoles and their separation from the cytostomal area by contractile influences have been considered. The changes in shape, size, pH and the state of degradation of food contents in DVI, II and III, culminating in the collapse of the egestion vacuole, have been described. The importance of recycling vacuole membrane in the form of discoidal vesicles which assemble in the cytostome-cytopharynx area to coalesce with a growing vacuole, has been emphasized. Details of feeding organelles are considered in later chapters where appropriate.

7 METABOLIC PATHWAYS

The previous two chapters have traced some of the ways by which heterotrophic protozoa obtain their nutrients. Once inside the organism these food components must first be broken down via catabolic pathways, then they may be reassembled as storage compounds or be used in various biosyntheses required for cell growth. The multiplicity of biochemical pathways and reactions dictates that this chapter can do no more than approach the subject in a most superficial way. A reader who wishes to consolidate his knowledge or go further into the study of metabolic pathways should consult a comprehensive biochemistry text.

The organic requirements of protozoa are, not unexpectedly, linked to their methods of nutrition. Autotrophic flagellates are the least demanding, being satisfied with inorganic or simple organic carbon compounds. Amoebae have moderate requirements and ciliates, parasitic flagellates and apicomplexans have very precise requirements for organic nutrients for energy production and protein synthesis. The relatively small number of protozoa which have been grown in chemically defined media illustrates the complexity of the problem.

Carbohydrates, proteins and lipids, as the major organic food components, are discussed in relation to protozoan metabolism. It will become apparent that much more information is available on the metabolic pathways involving carbohydrates than on lipids and proteins.

Carbohydrate Metabolism

Most protozoa, with the exception of a small number of blood parasites, store carbohydrate in some form of polysaccharide, but their choice of polymer is not necessarily that of the majority of animals, glycogen. It may be starch, amylopectin, glycogen or the β 1,3-linked polymers, paramylon and leucosin. Bacteria, in contrast, synthesize polyhydroxybutyrate which is metabolically and structurally similar to lipids, as their important carbon source.

Carbohydrate Storage

All carbohydrate reserves are polymers of glucose assembled in different ways, but the choice of storage material is to some extent linked to the particular class of protozoa, or even to the genus. It may be any one of the carbohydrates listed above and in Table 7.1. The identification of amylopectin as a reserve polysaccharide has revealed that this substance is more widespread than was originally suspected. 'Coccidien glykogen' of the *Eimeria* species has recently been identified as amylopectin, present in granules 0.5 by 0.7 μm (Ryley *et al.*, 1969; Wang *et al.*, 1975).

Table 7.1: Carbohydrate Reserves in Selected Protozoan Orders.

	Food Reserve		Representative
	β-1,3-glucan	α-1,4-glucan	Species
Mastigophora:			
Chrysomonadida	leucosin chrysolaminarin		*Ochromonas*
Cryptomonadida		starch	*Cryptomonas*
Dinoflagellida		starch	*Ceratium*
Prymnesiida	leucosin		*Chrysochromulina*
Euglenida	paramylon		*Euglena*
Volvocida		starch	*Chlamydomonas*
Chloromonadida		glycogen?	*Vacuolaria*
Trichomonadida		glycogen	*Trichomonas*
Hypermastigida		glycogen	*Trichonympha*
Sarcodina:			
Amoebida		glycogen	*Amoeba, Entamoeba*
Apicomplexa:			
Eucoccidiida		amylopectin	*Eimeria*
Piroplasmida		glycogen?	*Babesia*
Myxozoa:			
Bivalvulida		glycogen	*Henneguya*
Microspora:			
Microsporida		glycogen	*Nosema*
Ciliophora:			
Trichostomatida		amylopectin	*Isotricha, Dasytricha*
Entodiniomorphida		amylopectin	*Entodinium,Epidinium*
Hypotrichida		glycogen	*Euplotes*
Hymenostomatida		glycogen	*Paramecium*

Most ciliates and generally also the sarcodines, store glycogen or paraglycogen in the form of single spherical granules or as rosettes. The parasitic *Entamoeba histolytica* adopts the rosette form. The type of glycogen stored by a particular ciliate is determined by its generic relationship and not by different food preferences or habitats. For instance, *Euplotes crassus*, a marine species, and *E. eurystomus*, a freshwater species, both feed on algae and store α glycogen (the rosette form). Two freshwater species of *Oxytricha*, *O. bifaria* which feeds on bacteria and *O. fallax* which feeds on algae, both store paraglycogen (Verni and Rosati, 1980). Storage of glycogen in discrete granules is the most common pattern amongst ciliates, exceptions being the rosette form in *Euplotes* and a diffuse distribution found in *Paraurolnema* (Kareem and Soldo, 1978).

The most notable exceptions in the choice of carbohydrate reserve among the ciliates is that of the ciliates of ruminants which lay down considerable amounts of amylopectin in the form of small granules dispersed through the cytoplasm (Figure 8.6, and Oxford, 1955). The metabolic behaviour of this group of ciliates is discussed in Chapter 8.

Flagellates vary considerably in their choice of carbohydrate. This is consistent with the greater range of nutritional types from photoautotrophs to obligate heterotrophs. Table 7.1 shows the type of carbohydrate reserves in some examples of the protozoan orders, with the Mastigophora showing the biggest variation. In the Zoomastigophora, such as the trichomonads, glycogen is stored, while in trypanosomes no significant amounts of reserve polysaccharide have been detected as storage of substrate is largely unnecessary in parasites of the bloodstream, an extremely substrate-rich medium. The parasitic apicomplexa show a similar pattern. *Plasmodium* resembles the trypanosomes in having no need of reserves, while the gregarines and *Eimeria* store amylopectin in similar fashion to the rumen ciliate protozoa.

Sources of Carbohydrate

Carbohydrates enter protozoa in a variety of ways that do not necessarily involve the ingestion of polysaccharide-containing solid food, which must first be broken down into simpler molecules before utilization. Many protozoa absorb simpler sugars routinely and by chance during phagocytosis and pinocytosis when they ingest fluid medium into endocytotic vacuoles. Free-living amoebae take in glucose and sucrose by pinocytosis very rapidly in the presence of inducers such as albumin and Alcian Blue and during treatment with calcium ionophore (Chapman-Andresen and Holter, 1955; Prusch, 1980). Parasitic *Entamoeba histolytica* takes in only a very small proportion of its glucose requirement by endocytosis. It possesses a specific membrane-mediated transport system for glucose uptake which accounts for 1,000 times more sugar than by endocytosis alone (Serrano and Reeves, 1975).

Parasitic flagellates, species of *Trypanosoma*, use membrane-mediated mechanisms for the uptake of sugars. The existence of specific sites has been demonstrated. Ruff and Read (1974) showed that *T. equiperdum* has a single hexose site used for the uptake of glucose, fructose and mannose. Arme (1982) reviews carbohydrate uptake in other parasitic flagellates. *T. lewisi* has two distinct membrane transport sites which take up glucose, mannose, fructose, galactose, glucosamine and 3-O-methyl glucose.

Cellulose as a Carbohydrate Source

Some ciliates in ruminants and some amoebae possess α-amylases (operative within the pH range 4.8-6.0) to break down plant hemicelluloses and starch to simpler sugars. Cellulase activity has been demonstrated in hartmanellid amoebae, in flagellates of termites and woodroaches, and in entodiniomorphid ciliates of ruminants, although the possibility that in some cases this activity may be due to cellulolytic bacteria cannot be ruled out with certainty (Coleman *et al*., 1976; Hungate, 1955). Only by culturing more species axenically will it be possible to prove that cellulose breakdown is achieved without the aid of bacteria. *Trichomitopsis* from termites grown axenically in anaerobic conditions contains no endosymbiotic bacteria and produces its own cellulase (Yamin, 1978).

Energy Production

The initial breakdown of ingested carbohydrates involves standard hydrolytic pathways, although protozoa usually contain more elaborate sets of enzymes than many animal cells. *Tetrahymena,* for example, is able to hydrolyze maltose, isomaltose, cellobiose, sucrose and starch.

Energy production by oxidative metabolism of carbohydrates has been demonstrated to operate in protozoa as in other organisms, and to involve the three main phases, starting with the conversion of sugars to pyruvate by the Embden-Meyerhof glycolytic pathway. Further metabolism of pyruvate continues aerobically in the tricarboxylic acid cycle, from which set of reactions surplus hydrogen atoms pass into the electron transport chain through respiratory pigments such as the cytochromes. An alternative route, the pentose phosphate pathway or phosphogluconate pathway, for the catabolism of sugars, operates in many protozoa as in other animals.

In protozoa metabolic pathways involving carbohydrates all vary in detail, so presenting a comprehensive scheme is an impossible task. A simple outline scheme which may be used for reference seems the most appropriate at this stage, followed by detailed consideration in the relevant chapters.

Embden-Meyerhof Glycolytic Pathway

Figure 7.1: Embden-Meyerhof Glycolytic Pathway.

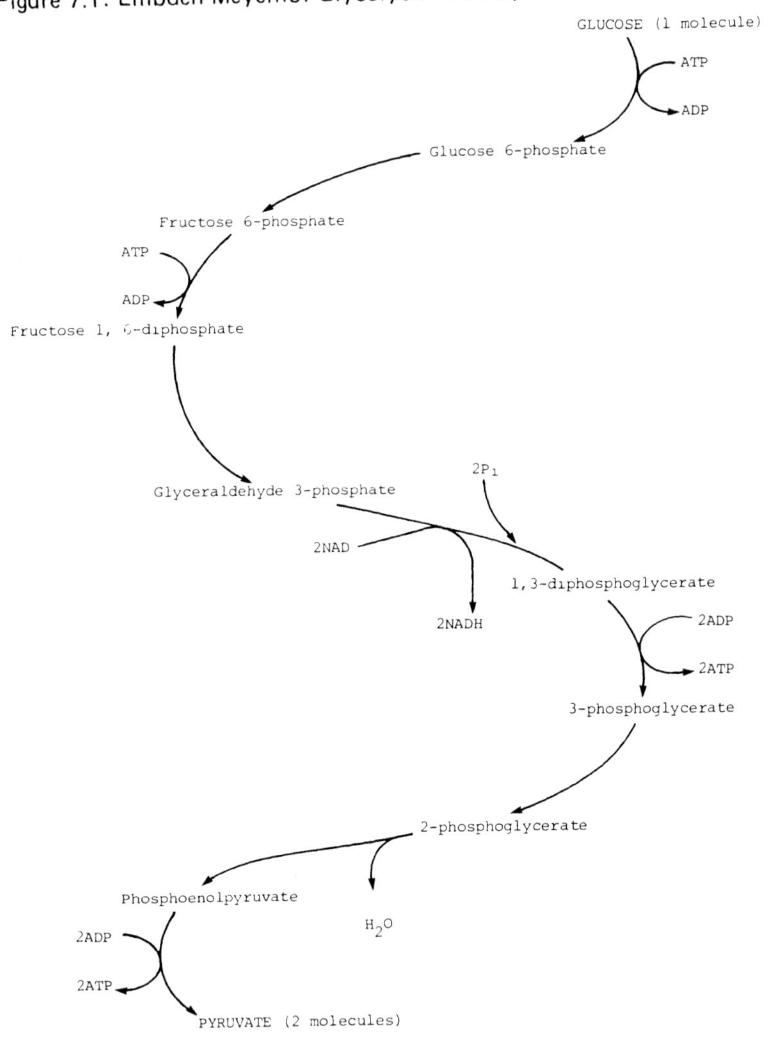

The starting point of the Embden-Meyerhof glycolytic pathway is glucose or, less directly, glycogen. Each six-carbon glucose molecule passes through a series of degradative steps to become two three-carbon pyruvate molecules with a net gain of two energy-rich ATP molecules (Figure 7.1). Two molecules of ATP are required to provide energy for the initial steps of the glycolytic pathway, one to phosphorylate glucose and make possible the conversion to a fructose configuration and a second one to add a further phosphate group, giving fructose-1,6-diphosphate. By splitting this fructose compound two molecules of glyceraldehyde-3-phosphate are produced which are each able to receive a second phosphate group to give 1,3-diphosphoglycerate. The consequence of adding second phosphate groups here is to release hydrogen which is picked up by the hydrogen-carrying coenzyme, nicotinamide adenine dinucleotide (NAD). NADH so produced transports its hydrogen atoms for use in reductions later in the glycolytic pathway. The conversion of energy-rich 1,3-diphosphoglycerate to 3-phosphoglycerate involves the release of a phosphate group which is used to synthesize ATP from ADP. A change in configuration from 3-phosphoglycerate to 2-phosphoglycerate makes possible the production of phosphoenolpyruvate by dehydration. Phosphoenolpyruvate is a high-energy compound which readily releases its phosphate group to ADP, leaving pyruvate, which is the standard end product of the Embden-Meyerhof pathway, for transfer to the tricarboxylic acid cycle.

Pyruvate is the most common end product in the glycolytic pathway. The kinetoplastid flagellate *Trypanosoma brucei* and related species in the slender trypomastigote form, have a standard glucose catabolism pathway, producing mainly pyruvate, but this product is not apparently incorporated into the tricarboxylic acid cycle for further degradation. The enzymes required for the TCA cycle and the cytochrome system have not been demonstrated, nor does the mitochondrion itself contain the well-developed cristae characteristic of an active organelle. These two facts suggest that mitochondrial respiration does not occur in the bloodstream phase of *T. brucei*, that enough energy is produced during the initial degradation of glucose and that further degradation is superfluous. For parasites living in oxygenated substrate-rich blood, this abbreviated metabolic pathway is adequate (Bowman and Flynn, 1976; Flynn and Bowman, 1973; Ryley, 1956).

Tricarboxylic Acid Cycle

Returning to the main theme of the pathways of energy flow, pyruvate produced by the phosphorylation and breakdown of glucose is incorporated into the TCA cycle, the enzymes for this being located in the mitochondria. Before entering the TCA cycle, pyruvate must first be decarboxylated to acetyl Co-A, losing hydrogen and carbon dioxide during the reaction, but making possible its subsequent combination with oxaloacetic acid (Figure 7.2). The involvement of the TCA cycle in oxidative matabolism is not restricted to carbohydrates (through pyruvate) for fatty acids and amino acids are also converted to acetyl Co-A and oxidized by this pathway. As the TCA cycle is well known it would be superfluous to describe it here in much detail. An outline is shown in Figure 7.2, indicating the principal oxidations and transformations. The starting point of the TCA cycle is acetyl Co-A, from decarboxylated pyruvate, which combines with oxaloacetic acid to form citric acid. After two enzyme-mediated decarboxylations oxaloacetic acid is reformed.

Figure 7.2: The Tricarboxylic Acid Cycle

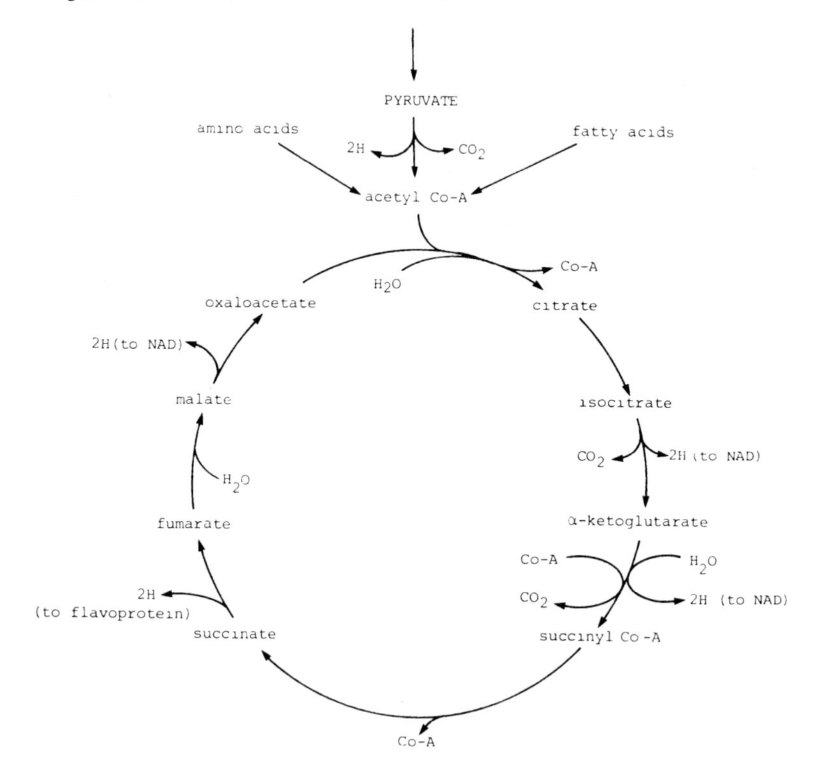

Many of the TCA cycle enzymes have been identified in protozoa, except in bloodstream trypanosomes which have already been mentioned as not having well-developed cristae in their mitochondria.

Related kinetoplastid flagellates, the leishmanias, have a classical TCA cycle, both as amastigotes in macrophages of mammalian blood and as promastigotes in the gut of insect vectors (Chapter 9). Hart and Coombs (1982) have shown that the promastigotes of *Leishmania mexicana* channel carbohydrates (after glycolysis) and amino acids, and probably fats (after β-oxidation) through the TCA cycle to the terminal respiratory pathway (electron transport chain). It is perhaps surprising that the pathways of catabolism and energy production appear similar in leishmanias in both stages of the life cycle when their alternative hosts provide such dissimilar environments of nutrient status and temperature.

The Electron Transport Chain

This final stage in the oxidation of glucose after all the carbon atoms have been oxidized to carbon dioxide, has two purposes to achieve. One is to pass hydrogen atoms, mainly those surplus from the TCA cycle, through a series of reactions called the cytochrome system to release them as water and simultaneously to re-oxidize NADH (the hydrogen carrier). The second function linked with this is the creation of energy-rich bonds by the oxidative phosphorylation of ADP to ATP.

The cytochrome system or respiratory chain is located in the mitochondria. The starting point is flavoprotein which receives hydrogen atoms from succinic acid. These hydrogen atoms release electrons for passage through a chain of respiratory pigments, cytochromes b, c and a_3, which may contain iron in the oxidized (Fe^{3+}) and reduced (Fe^{2+}) states, a convenient electron transport system ending in the production of water. In balancing the energy equation it can be shown that for every pair of hydrogen atoms oxidized to water, three molecules of ADP are phosphorylated to three molecules of energy-rich ATP by a complex multi-enzyme system. The energy locked up in ATP has a crucial part to play in the initial breakdown of glucose molecules which occurs in the glycolytic pathway illustrated in Figure 7.1.

The metabolic pathways for the utilization of carbohydrate in energy production, which have been outlined above, operate in an aerobic environment. Many protozoa, including parasites of the body cavity and the gut, live in an atmosphere with little or no oxygen. Substrate utilization and energy production in these species involve principally anaerobic fermentations, although some parasitic amoebae such as *Entamoeba histolytica* have the enzymes to catabolize glucose aerobically also (Gutteridge and Coombs, 1977). The parasitic flagellates, *Trichomonas* and *Giardia*, the hypermastigid flagellates which digest cellulose in wood-eating insects, and the ciliates of ruminants, are all considered to be anaerobic organisms with little tolerance for oxygen (von Brand, 1973; Honigberg, 1967; Hungate, 1939, 1943a, 1978; Lindmark, 1980).

Pathways of anaerobic catabolism yield a variety of end products through pyruvate to lactate, succinate, acetate, and carbon dioxide and water. Metabolic end products in relation to certain protozoan species will be considered further in the following chapters.

Acetate figures frequently as an end product in anaerobic catabolism. It is also important as a precursor for carbohydrate storage in photoautotrophic flagellates (*Euglena*) and in the ciliate *Tetrahymena* (Hogg and Kornberg, 1963; Smillie, 1968).

Protein Metabolism

Amino acids are the essential building bricks for protein synthesis in protozoa, as they are in all animal cells. A limited number of photoautotrophic flagellates have the capacity to synthesize their own amino acids using inorganic sources of nitrogen such as nitrate and ammonia. Others require a supply of essential amino acids in their growth medium. The quest for chemically defined growth media has resulted in detailed information on amino-acid and specific growth factor requirements. *Euglena gracilis*, as one of the first protozoa to be grown in a chemically defined medium, has attracted the attention of many researchers. Much of the information on cultivation and growth requirements of this organism has been summarized by Cook (1968). Among the ciliates *Tetrahymena pyriformis* became the favoured organism for culture studies. It requires a growth medium containing sixteen amino acids, uracil, adenylic, cytidylic and guanylic acids, acetate, glucose, a purine, a pyrimidine, vitamins and mineral salts (Hill, 1972; Kidder and Dewey, 1951). Other *Tetrahymena* species have been grown in variations of this medium and even with fewer amino acids. Four species of *Paramecium*, *P. calkinsi*, *P. caudatum*, *P. aurelia* and *P. multimicronucleatum*, can be maintained continuously on a medium with an amino-acid content based on the composition of casein, five nucleotides, sodium acetate, oleic acid, stigmasterol, vitamins, salts and trace metals. *Acanthamoeba* can be grown in a medium consisting of eighteen amino acids, vitamin B_{12}, thiamin, acetate, citric acid, salts and trace metals (Adam, 1959; Byers *et al.*, 1980). A summary of chemically defined media for the cultivation of some parasitic protozoa may be found in Gutteridge and Coombs (1977).

Heterotrophic protozoa require a range of readily available amino acids which are essential for protein synthesis. Twenty different amino acids have been identified in the proteins of animals and plants and the list in Table 7.2 is familiar to all students of biochemistry and physiology.

Table 7.2: Amino Acids Required for Protein Synthesis Grouped into 'Families'.

(Glutamate)	Glycine
Glutamine	Cysteine
Arginine*	Serine*
Proline	
	Alanine
(Aspartate)	Valine*
Asparagine	Leucine*
Methionine*	
Threonine*	Histidine*
Isoleucine*	
Lysine*	
Phenylalanine*	
Tyrosine	
Tryptophan*	

Note: * Essential for growth; the remainder can be synthesized *de novo* or from other amino acids.

Sources of Amino Acids

Because of the wide range of metabolic behaviour in free-living and parasitic protozoa, it is not possible to produce a definitive list of amino acids essential for protein synthesis in this group. Of the twenty amino acids listed in Table 7.2, eleven are identified as being commonly required from exogenous sources while the others can be synthesized within the protozoon if the appropriate precursor is present. Synthesis would normally be from some other amino acid rather than from nitrate and ammonia. Exogenous amino acids required by *Tetrahymena* are arginine, histidine, isoleucine, leucine, lysine, methionine, phenylalanine, threonine, tryptophan and valine, although a certain amount of compensation is possible (Wu and Hogg, 1956). For instance, the absolute requirement for phenylalanine may be reduced by 50 per cent if the closely-related tyrosine is present.

Amino acids enter the protozoan body as protein ingested during endocytosis, although many parasitic protozoa acquire amino acids directly from the host body fluids by simple diffusion or by membrane-mediated transport, often at specific sites. Proteins taken in by endocytosis are hydrolyzed within the endocytotic vacuole and are broken down to constituent amino acids for rebuilding body protein, nucleic acids and enzymes. Many of the proteases and peptidases involved in the hydrolysis of proteins are specific to protozoa and differ from those found in mammalian cells. Most operate in an acid environment, with an optimum range between pH 3.7 and 5.7, similar to that required for the hydrolysis of carbohydrates in digestive vacuoles. However, maximum proteolytic enzyme activity in some protozoa is outside this range; *Entamoeba histolytica* peaks at pH 7.0 and the rumen ciliate *Entodinium caudatum* at pH 6.5-7.0 (Abou Akkadar and Howard, 1962).

Lipid Metabolism

Investigations into the lipid metabolism of protozoa have lagged far behind those of carbohydrate and protein metabolism. The lipid requirements of the majority of protozoa studied are known to include some of a large range of saturated and unsaturated fatty acids and one or more sterols, the most common being cholesterol, ergosterol and stigmasterol. Sterols are important components of all eukaryotic organisms and their special function in protozoa is in helping to maintain the stability of the plasma membrane. The total lipid content of protozoa varies from species to species and some idea of the range is shown in Table 7.3.

Many phytoflagellates can synthesize their own metabolically important lipids. This lipid content is relatively high, as chloroplasts contain large amounts of essential sulpholipids. The chrysomonad, *Ochromonas*, is independent of external lipid supplies, being able to synthesize four sulpholipids and arachidonic acid. Although no exogenous sterol requirement has been demonstrated in *Euglena gracilis*, ergosterol has been isolated from this species, indicating that the flagellate does indeed have the appropriate metabolic pathways to synthesize its own sterols. Other phytoflagellates which are colourless and have evolved as heterotrophs rather than pigmented autotrophs, require lipids from their environment. The heterotrophic dinoflagellate *Oxyrrhis marina* must have an exogenous source of lipids (Droop, 1954, 1959), and *Peranema*, a colourless euglenoid, grows well on dried yeast in which the major sterol is ergosterol.

Table 7.3: Lipid Content of some Protozoa.

Organism	Lipid Content as % of Dry Weight
Phytomastigophorea:	
Euglena gracilis	27
Ochromonas danica	35
O. malhamensis	33
Zoomastigophorea:	
Trypanosoma rhodesiense	11
T. brucei	16
T. lewisi	12
Ciliophora:	
Tetrahymena pyriformis	24
Apicomplexa:	
Plasmodium knowlesi	29
Eimeria acervulina (oocysts)	14

Source: Aaronson and Baker, 1961; Gutteridge and Coombs, 1977.

Zooflagellates present a more complex picture both in absolute lipid requirements and in synthetic abilities. The most studied zooflagellates are the trypanosomes, but even within this genus considerable variations occur. Trypanosomes behave differently in different hosts and in culture *in vitro*. Most trypanosomes can synthesize sterols, usually ergosterol, but the bloodstream form of *T. rhodesiense*, with inactive mitochondria, must be supplied with a sterol, cholesterol, which is available in the blood. The trypanosomatid *Herpetomonas samuelpessoai*, isolated from its insect host, can also synthesize ergosterol from acetate, or from methionine in the absence of acetate (Fagundes *et al.*, 1980). It is possible that trypanosomes can also synthesize fatty acids, although this has not yet been demonstrated. Other parasitic flagellates have more restricted synthetic abilities than trypanosomes. Trichomonads generally require cholesterol or related sterols and certain fatty acids, although *Trichomonas vaginalis* can synthesize sterol in the presence of acetate (Roitman *et al.*, 1978). Trichomonads are unable to synthesize the C_{14} to C_{18} range of saturated and C_{18} to C_{22} unsaturated fatty acids (Shorb and Lund, 1959).

Ciliates require sterols and fatty acids in clearly defined proportions, otherwise deleterious effects appear. *Paramecium aurelia* needs oleic acid, although concentrations over 10 μg/litre actually inhibit growth. This inhibition is partially annulled either by the presence of certain non-essential fatty acids such as those in Tween 60, whose major component is stearic acid, or by increasing the concentration of stigmasterol which is required for optimal growth (Soldo and von Wagtendonk, 1967). *Paramecium*, particularly in the young log-phase stage, takes up lipids from the external medium and stores them in cytoplasmic vesicles until required (Kaneshiro *et al.*, 1979). One of the major uses of lipids in these organisms is as a component of the membrane covering the cell and especially the ciliary membrane, which contains a high proportion of unsaturated fatty acids. The alternative use of diverting lipids, after β-oxidation to shorter chain fatty acids, to the TCA cycle has already been mentioned. Among the ciliates *Tetrahymena pyriformis* has the capacity to synthesize its own lipids and so is independent of exogenous sources (Koroly and Conner, 1976). The polycyclic alcohol tetrahymenol, synthesized by *Tetrahymena* as a prominent constituent of its ciliary membrane, actively enhances cell growth when added to the culture medium (Conner *et al.*, 1982). Surprisingly perhaps, lipid-synthesizing ability is not a generic character, for *T. corlissi*, *T. paravorax* and *T. setifera* all require a sterol, preferably cholesterol. Without cholesterol, growth of *T. paravorax* is poor and its absence cannot be adequately compensated for by the addition of precursors of cholesterol.

The pathways of lipid metabolism in protozoa are complex and variable. The uses of lipids are more obvious. The degradation of long chain fatty acids by β-oxidation to acetyl Co-A and the TCA cycle has already been mentioned. However, as a source of energy for protozoa this pathway is certainly of much less importance than that of carbohydrate glycolysis, as most protozoa store little lipid. The main concentration of lipids is as membrane constituents, particularly in the surface membrane of organelles of movement, cilia and flagella. Considering the constancy of structure internally of cilia and flagella, it may be surprising to find that the lipid constituents of their surface membranes vary considerably. The ciliary membrane of ciliates such as *Paramecium* and *Tetrahymena* contains high proportions of phosphonolipids and sphingolipids, whereas the flagellar membrane of the phytoflagellate *Ochromonas* is rich in sulpholipids. These differences may be correlated with different evolutionary pathways within the protozoa, but the end result is the same, a surface membrane which resists degradation (Rhoads and Kaneshiro, 1979).

In Summary

Carbohydrate metabolism in most protozoa results in the synthesis of storage polysaccharide, starch, amylopectin, glycogen, paramylon or leucosin. Notable exceptions are certain blood parasites, *Trypanosoma* and *Plasmodium*, which appear not to need such reserves. Carbohydrates enter the organism by endocytosis during feeding or by membrane-mediated transport, often at specific sites.

The standard metabolic pathways, the Embden-Meyerhof glycolytic pathway and the TCA cycle have been described, followed by the cytochrome-mediated electron transport chain. In anaerobic conditions metabolic end products are fatty acids, particularly lactate and acetate.

Protein metabolism requires a range of 20 amino acids; eleven must be supplied from exogenous sources in heterotrophic protozoa. Some ciliates can synthesize their own sterols, but most need at least one exogenous source, which is used, in part, towards producing membrane for the cilia and the body of the organism. Protozoa have a relatively low lipid content.

8 FEEDING IN THE CILIOPHORA

Most of the preceding chapters have taken a rather generalized approach towards the process of feeding in protozoa, considering first the types of habitats and how these influence the food sources available in them. Then the endocytotic vacuole was presented as a dynamic unit of digestion, followed by a review of the metabolic pathways through which food once inside the organism could be utilized. Appropriately now a more specific approach is required, so that the particular characteristics within groups can be highlighted. For a variety of reasons some groups of protozoa have attracted more attention than others. The ciliates which live in the ruminant fore-stomach are of interest to agricultural researchers because these micro-organisms have a part to play in the digestive processes of sheep and cows. Ciliates which can be maintained on chemically-defined media are important in biochemical and physiological studies. Because there are so many interesting groups of ciliates, selection for special mention is difficult.

As a phylum the Ciliophora have developed a complexity of structure not found in other protozoa. In addition to the obvious character of possessing cilia, they show nuclear dualism, macronuclei and micronuclei with different functions, and their own version of sexual reproduction, by conjugation. Ciliates are a fully heterotrophic group in which the major food source is solid food taken in at a fixed point, the cytostome, except in a small group of astomatous ciliates which live in the body cavities of other animals. A description of food-trapping in ciliates necessitates using specific terms which should be defined to avoid confusion. They have been kept to a minimum here, but a full list of terms and concepts can be found in Corliss (1979).

Ciliate Organization

By definition the ciliates may be expected to have cilia at some stage in their life history. The structure of a simple cilium with its central axoneme of microtubules, nine doublets and a central pair, is familiar to all students of biology. The cilium arises from a kinetosome, a generative body below the pellicle or outer covering of the organism. Kinetosomes and their associated network of microtubules and microfilaments which make up the infraciliature have been illustrated in Figure 1.1. The infraciliature of a species is described by Corliss (1979) as 'a very conservative and fundamental feature', thus recognizing its diagnostic contribution.

Ciliary Organelles

Cilia become grouped into elaborate organelles which may perform a double duty, movement and food-trapping. Perhaps the simplest grouping of cilia is into cirri, tufts of up to 100 cilia joined by their kinetosomes, used as 'walking legs' or for food-gathering in the hypotrichs (Figure 8.1(a)). Cilia arranged in linear series, usually beating metachronally, and with their kinetosomes closely set, are described as paroral (undulating) membranes or as membranelles of various kinds. Inevitably confusion arises over nomenclature. Membranes and membranelles differ in the number of parallel rows of cilia involved.

A paroral (undulating) membrane, usually within or surrounding the feeding area, arises from a double row of kinetosomes where one row is barren and the other bears cilia (Figure 8.1(b)). A haplokinety is an alternative name for this ciliary membrane. A ciliary membrane is a compound structure often consisting of a series of three or more cilia with their kinetosomes so close set that they behave as a single but more effective cilium (Figure 8.1(c)). Adoral zone membranelles (AZM) organized on this pattern are important feeding and locomotory organelles. The quadrulus and peniculus in the buccal cavity of *Paramecium* are ciliary membranelles of this type, although consisting of four or more rows of cilia.

Ciliate Oral Cavities

Food normally enters the organism through a well-defined cytostome which may be superficial or lie deep in the base of a concavity. The cytostome is the oral opening, a special point in the pellicle which consists of plasma membrane only and no underlying alveolar sacs. Here the endocytotic vacuole invaginates, taking with it any food which has been trapped by the feeding activities of the ciliate. It is usually associated with a cytopharynx when the two may be virtually inseparable.

In the more primitive ciliates the cytostome is apical or lateral, at the base of a short tubular section whose walls are strengthened by nematodesmata (bundles of microtubules). In others the cytostome is reached by a concavity variously called a buccal cavity, a vestibule or a peristomial cavity, which connects the cytostome-cytopharynx to the

Figure 8.1: Cilia Arranged into a Variety of Complex Organelles. (a) The cirrus of a hypotrich contains up to 100 cilia joined by their kinetosomes to form a 'walking leg'. (b) A paroral (undulating) membrane usually lies within or around the feeding area. It consists of a single row of cilia with two rows of kinetosomes, one row barren and one row cilia-bearing. (c) A ciliary membranelle consists of rows of three or more close-set cilia arranged to operate as single units of a linear series. These combined cilia produce a more effective beat.

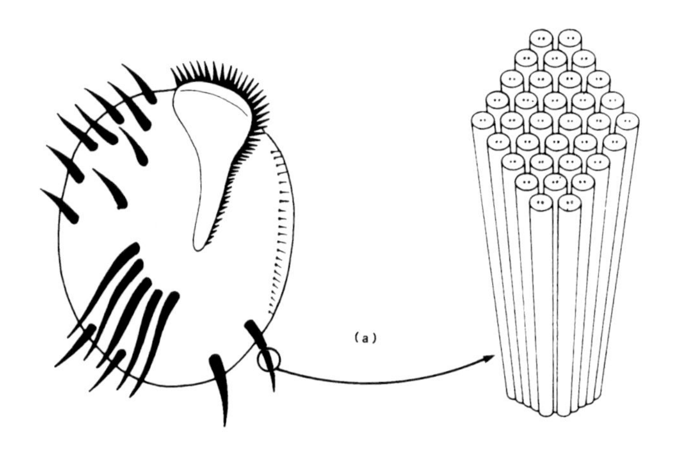

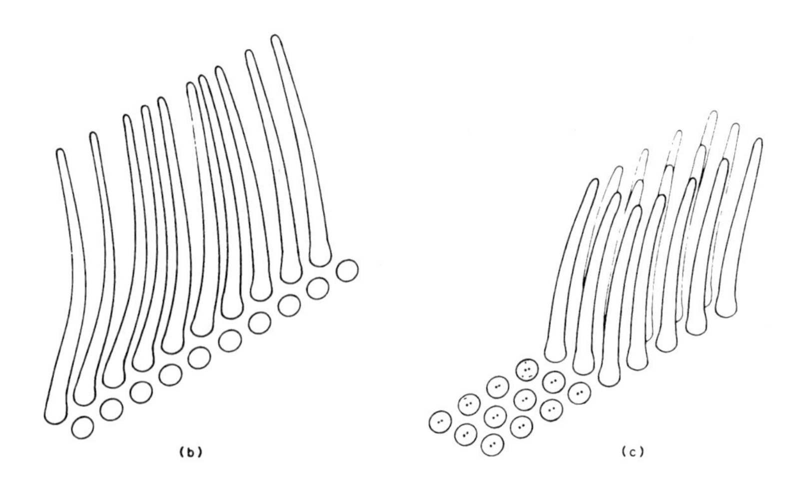

exterior of the animal. Nomenclature of the parts of the feeding apparatus remains confusing and inconsistent in spite of Corliss' (1979) attempt to unify it.

Principles of Classification

Ciliates have proved worthy organisms for ultrastructural studies, particularly in relation to the organelles of feeding. The large gaps which existed between the protozoologists' interpretation of structure and function are beginning to be filled, although at the same time others are being uncovered. The result of this increase in information, particularly on the oral ciliature and the oral apparatus of ciliates is that the classification of the Ciliophora has been revised to take account of the relationships and differences which were either not previously recognized or were not thought to be of fundamental importance. The erection of three new classes, Kinetofragminophorea, Oligohymenophorea and Polyhymenophorea regroups ciliates according to Corliss (1979), using behaviour of mouthparts at stomatogenesis and the relationship between the oral ciliature and somatic ciliature as some of the principal distinguishing characters. Stomatogenesis, or the production of new oral structures at division, involves the assembly of kinetosomes at the appropriate sites and in the appropriate arrangement where they can generate new oral ciliature in the daughter organisms.

The origin of these sets of oral kinetosomes indicates important evolutionary links not previously recognized (Corliss, 1973). In the Kinetofragminophorea stomatogenesis is generally telokinetal, where cilia arise from the kinetosomes of anterior somatic kineties (rows of cilia and their infraciliature or kinetofragments). This class contains the more primitive ciliates (Corliss, 1974; Puytorac *et al.*, 1974).

The Oligohymenophorea show a more complex and variable stomatogenesis. The new oral apparatus is generated either from kinetosomes derived by division of the parent buccal apparatus (buccokinetal) or from kinetosomes of somatic cilia lying posterior to the parent buccal cavity (parakinetal). The Polyhymenophorea contains the larger more robust ciliates with stomatogenesis, which is either parakinetal or apokinetal (kinetosomes appearing without any apparent connection with somatic or buccal cilia, *de novo*). A well-illustrated description of apokinetal stomatogenesis in *Euplotes* may be found in Tuffrau *et al.* (1976).

Organelles for Food-trapping

The oral apparatus which differentiates in the daughter organisms after division may be simple, furnished with cilia hardly distinguishable from somatic cilia on the body surface, or it may be an elaborate buccal cavity with conspicuous ciliary membranelles. Certainly the form of feeding organelle will determine whether the ciliate is a filter-feeder, a carnivore, a predator or a herbivore.

Filter-feeders

Many ciliates are filter-feeders, particularly those included in the Oligohymenophorea and Polyhymenophorea, and only a few can be selected for fuller description here. Some of the most efficient filter-feeders are the hymenostomes *Paramecium* (Allen, 1974) and *Tetrahymena* (Sattler and Staehelin, 1979) and *Glaucoma* (Fenchel and Small, 1980). All have a ventral buccal (oral) cavity containing well-developed ciliary membranelles and some form of paroral membrane on its right margin.

Paramecium buccal membranelles were first described by von Gelei (1934) and have since figured in many textbooks of protozoology. The arrangement of membranelles, the quadrulus, dorsal and ventral peniculi on the dorsal and left wall of the buccal cavity is well-known (Pitelka, 1969). The paroral membrane, a haplokinety with a double row of kinetosomes, one row bearing cilia and the other barren, still retains the name endoral membrane which was used in early descriptions. The right wall of the buccal cavity is lined by 40-50 longitudinal ribs which extend down towards the cytostome (Figure 8.2, and Allen, 1974). Rather surprisingly the ribbed wall lacks any supporting microtubules within the ribs as occur in other hymenostomes such as *Tetrahymena* (Sattler and Staehelin, 1979). Sets of microtubules do however feature as important conducting elements in the region of the cytostome-cytopharynx. One set of microtubules lies below the ribbed wall and extends beyond the right lip of the cytostome as the post-oral microtubules which guide the forming food vacuole into the cytoplasm. A second set, microtubule ribbons arising at the left lip of the cytostome, are believed to filter out and transport disc-shaped vesicles back to the left lip to coalesce with the expanding food vacuole (Allen, 1974). This form of membrane recycling has been discussed in Chapter 6.

Tetrahymena possesses ciliary membranelles which are homologous to those in *Paramecium*, although housed in a shallower oral (buccal)

Figure 8.2: A Reconstruction of the Buccal Cavity of *Paramecium*. (a) A left view shows the vestibule (ve) and ciliary membranelles on the left wall of the buccal cavity (bc), quadrulus (qu), dorsal peniculus (dp) and ventral peniculus (vp). (b) A right view shows the endoral membrane (em), 40-50 ribs (br) lining the right wall of the buccal cavity, and the right lip of the cytostome (rlc).

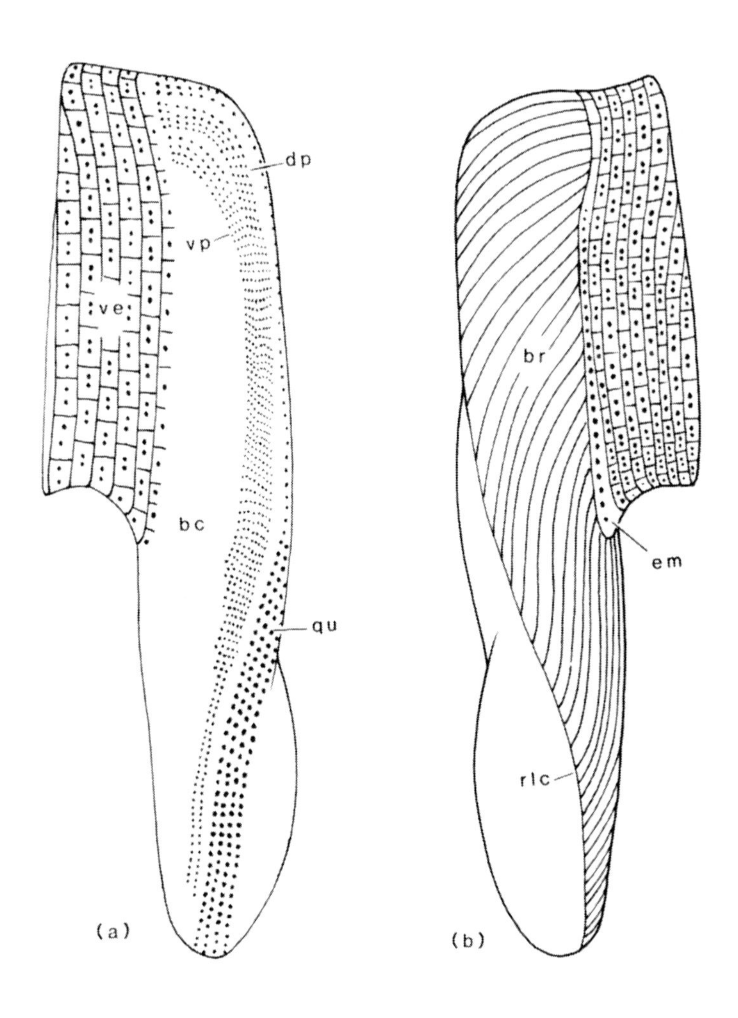

Source: Redrawn from Allen (1974).

cavity (Sattler and Staehelin, 1979; Williams and Bakowska, 1982). The three membranelles lying in the left side of the buccal cavity are polykineties consisting of three rows of cilia. The right wall of 17 ribs extending from the oral opening to the forming food vacuole at the cytostome, operates as a microslide down which food particles travel to the cytostome. Food particles are first trapped by the cilia of the undulating membrane, a haplokinety homologous with the endoral membrane of *Paramecium* on the right margin of the oral cavity. Then they are kept in motion by the sweeping action of the three membranelles (Figure 8.3).

Although the arrangement of ciliary organelles in *Tetrahymena*

Figure 8.3: Buccal Cavity and Feeding Organelles of *Tetrahymena,* in Ventral View. A co-ordinated sequence operates to trap food: first the undulating membrane (um) on the right margin of the buccal cavity, then ciliary membranelles (M1,2,3) on the left wall, finally 17 oral ribs (or) in the right wall slide food to the growing food vacuole (fv). See text for a fuller description.

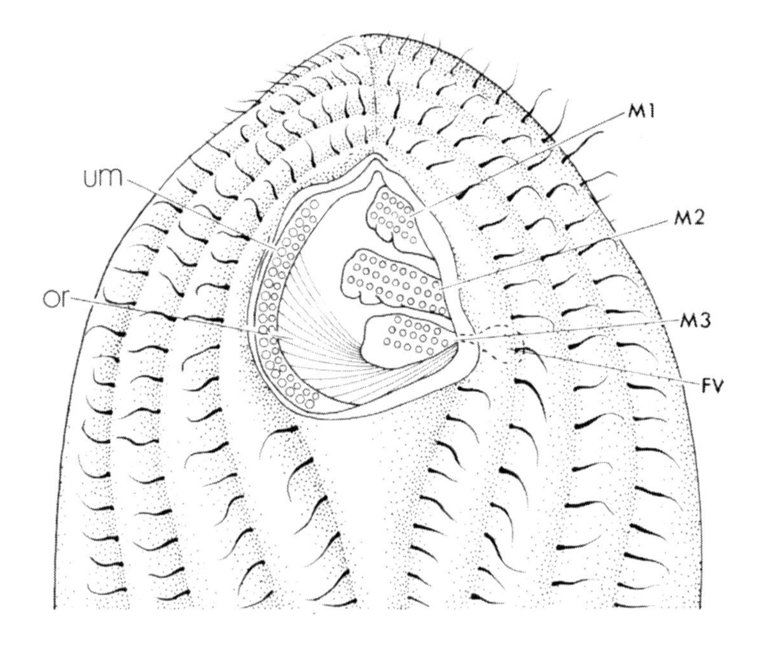

Source: Based on Sattler and Staehelin (1979).

resembles *Paramecium*, that of the microtubules is somewhat different. On the assumption that the 4 + 2 pattern of microtubules within the oral ribs in the 10 μm deep buccal cavity of *Tetrahymena* has a supporting function, then the absence of them in the larger buccal cavity of *Paramecium*, 25 μm deep, is difficult to explain (Figure 8.3).

Yet other hymenostomes such as *Cyclidium* and *Pleuronema* develop a sheet-like paroral membrane of fused cilia in three manoeuvrable sections which flicks rapidly out drawing in food particles in a fishing technique. The mode of action of the ciliary membranelles in the small hymenostome *Glaucoma* has been described in Chapter 5 and Figure 5.3. The size of the food particles caught depends on the distance between the cilia of the paroral membrane. *Cyclidium* does not easily trap particles of 0.23 μm or less, but *Glaucoma* retains particles down to 0.1 μm (Fenchel, 1980b).

The peritrich ciliates have developed filter-feeding to a fine art, as can be expected in a group which contains mainly sessile species. This efficiency is due to two characteristics in particular: the concentration of two bands of cilia on a raised peristome encircling the entire apical end and the habit of colonial or gregarious growth when the ciliary effects are additive. In *Vorticella* (Figure 5.3(a)) the efficient counter-clockwise rotary motion of the polykinety cilia results in particles moving past the ciliary tips at 2,500 μm s^{-1} (Sleigh and Barlow, 1976). The less active haplokinety traps incoming particles by forming a ledge with gaps between adjacent cilia of less than 1 μm.

A single *Vorticella* may extract particles from a water volume of 0.3 mm^3. Once trapped the particles are conducted round the peristome by the metachronal beat of the membranelles and down into the buccal cavity (Figure 6.2). The current-producing mechanisms which have been described in these few hymenostome and peritrich ciliates are adequate for the majority of bactivorous forms.

Herbivorous Ciliates

These ciliates have different problems from the filter-feeders. Two main problems have to be solved. First they must be able to accommodate food which is of an apparently unmanageable shape because of the cellulose walls when they ingest filamentous algae and secondly, they must have some means of drawing the food in. Many are hypostomes (Kinetofragminophorea), where the cytostome is on the ventral surface and is associated with elaborate strengthening devices of microtubules.

In *Nassula* the cytostome lies at the base of a shallow atrium. Internal to it a complex pharyngeal basket of rods and sheets of microtubules forms an internal support to the cytostome and probably plays an essential part in ingestion (Tucker, 1968). The breakdown of cyanobacterial filaments is initiated immediately they pass into the pharyngeal basket.

Also possessing a pharyngeal basket is the related genus *Pseudomicrothorax*, whose behaviour in ingesting cyanobacterial filaments has been described by Hausmann and Peck (1978, 1979) and Peck and Hausmann (1980). The pharyngeal basket in *Pseudomicrothorax* is tubular, appropriately shaped for the animal's food choice, and with the walls supported by 19-25 nematodesmata, each consisting of a bundle of approximately 150 microtubules (Figure 8.4 (e, f)). During feeding, lengths of filament are taken and are completely enclosed in a single giant endocytotic vacuole within 24 seconds of being drawn down the tubular pharyngeal basket. Even before the filament is completely enclosed by vacuole membrane, lysis of its innermost end is observed within five seconds of ingestion. To supply membrane for the rapidly growing vacuole and enzymes to break down the wall and lyse its contents, a stream of phagoplasmic vesicles passes forwards along the tubular basket to a perforated zone near its mouth.

Phascolodon has a similar type of ingestion organelle in which an arrangement of inwardly-projecting microtubule sidearms helps conduct food deeper into the organism. They are part of the pharyngeal basket which it uses to ingest diatoms (Figure 8.4(a)). At the top of the basket is a ring of tooth-like capitula whose main function is to pinch off the membranous invagination (endocytotic vacuole) as it forms below the cytostome, rather than actively to bite food organisms (Tucker, 1972). A cylindrical structure of this type functions as an efficient conducting mechanism for the newly formed food vacuole that is drawn down by the action of these inward projecting arms of the microtubules, which themselves have sufficient stability to withstand deformation during the passage of bulky endocytotic vacuoles. Elaborately structured ingestion organelles composed of rods or lamellae, ribbons or sheets of microtubules occur in many kinetofragminophoran ciliates. Rod-like nematodesmata are used to support the pharyngeal basket in the brackish-water ciliate *Chlamydodon* (Kaneda, 1960).

Herbivorous ciliates also occur amongst the heterotrichs (Polyhymenophorea). *Climacostomum*, a small relative of the much-studied *Stentor*, will feed on long filaments of green algae which it ingests whole, although it is mainly a filter-feeder on unicellular algae and

Figure 8.4: Feeding Organelles in Herbivorous Ciliates. (a) – (c), *Phascolodon*: (a) a pharyngeal basket (pb) is used to ingest diatoms; (b) the basket with a ring of tooth-like capitula (ca) to pinch off the endocytotic vacuole; (c) the ribs of the basket are nematodesmata (nm), bundles of microtubules with inwardly-projecting side arms (sa) performing a conducting role. (d) - (f), *Pseudomicrothorax*: (d) a tubular pharyngeal basket for ingesting cyanobacterial filaments; (e) and (f), its walls are supported by nematodesmata each of approximately 150 microtubules.

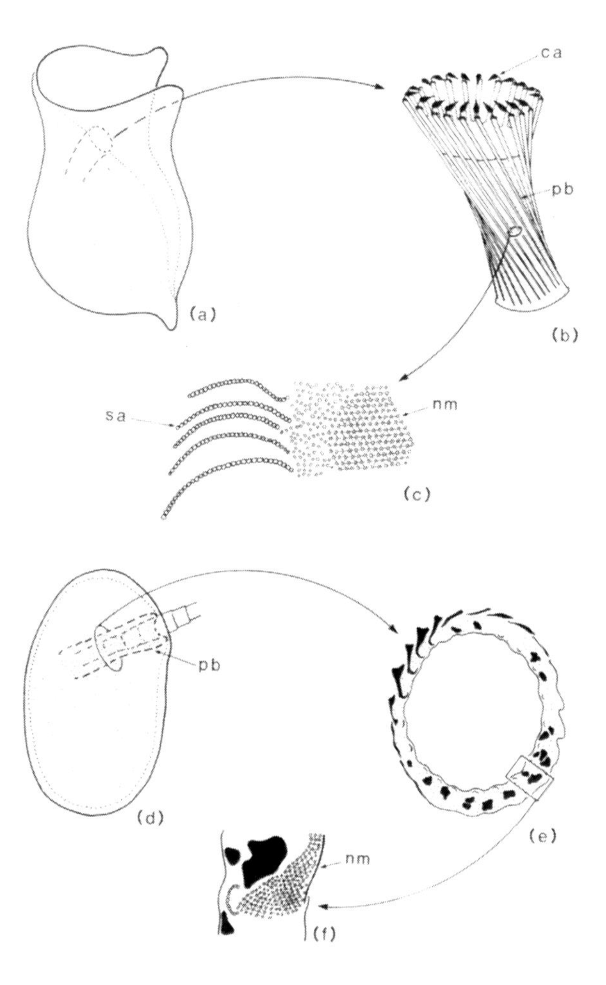

Source: *Phascolodon* (based on Tucker, 1972); *Pseudomicrothorax* (based on Peck and Hausmann, 1980).

other small organisms (Fischer-Defoy and Hausmann, 1977). Structurally its buccal organelles show similarities to hymenostomes and peritrichs, particularly the haplokinety which forms a collar round the cytostome. Characteristic constrictions of the buccal tube and body bending which occur during ingestion of large food may be helping to move the food downwards (Peck *et al*., 1975, and Figure 8.4).

Histophagous Ciliates

A number of ciliates respond to exudates from animal tissues and have exploited this response by becoming active scavengers. *Coleps* and *Prorodon* have developed the scavenging habit, although not to the total exclusion of other types of feeding. Sheet-like membranous organelles associated with feeding are a feature of apostome ciliates which live in or on a variety of animal hosts, arthropods, echinoderms and sea anemones (Grimes, 1976). The small rosette-shaped cytostome is used for taking in a mainly liquid diet by rapid pinocytosis (Bradbury, 1966, 1973). In *Hyalophysa* and *Polyspira*, exuviotrophic feeders on crustaceans, these pinocytotic vesicles coalesce to one large food mass and when this collapses peculiar sheet-like membranelles separate from it and return to the region of the cytostome. Similar production of sheet-like membranelles occurs in *Foettingeria* and *Ophiuraespira* apostomes which feed on the gut contents of anemones and ophiuroids.

Carnivorous Ciliates

A carnivore needs a more positive food-trapping mechanism than does a filter-feeder or a herbivore, and a means of responding to food contact. These responses operate through toxicysts, with paralytic and proteolytic secretions, and an extrusible or deformable proboscis-like structure, as most carnivores take food at least as big as themselves.

The initial contact may be due to chemotactic orientation or fortuitous contact. Chemotactic response is seen in the predators *Spathidium* and *Lionotus*, which are attracted to *Colpidium*, especially if damaged organisms are presented. Fortuitous contact between predator and prey occurs in the case of *Didinium* and *Paramecium* where both species swim in a random fashion so that feeding success is related to the density of paramecia (Wessenberg and Antipa, 1970). The proboscis of *Didinium* is furnished with two types of extrusosomes which eject on contact with a paramecium. Pexicysts attach to the surface of the prey and toxicysts penetrate up to 20 μm into it whilst remaining attached to the *Didinium* proboscis. Controlled cytoplasmic streaming down the proboscis draws the prey, attached by discharged toxicysts, into the cytostome-cytopharynx (Wessenberg and Antipa, 1970). Some loss of prey cytoplasmic fluid results when the prey is compressed for ingestion, and finally the cytostome closes up again.

Sorogena has unique feeding behaviour, but is considered to be structurally sufficiently related to *Didinium* and the other rapacious carnivores in the Haptorida (Kinetofragminophorea) to be classified with them. This new genus recently described by Bradbury and Olive (1980) feeds avidly on *Colpoda* until the prey is depleted. Numbers of the predator then aggregate on the bacterial film at the surface of the water into a sorocarp, a resting stage so called because of its resemblance to fruiting bodies of the slime moulds. When feeding, an anterior deformable mouth surrounded by a raised lip allows it to ingest *Colpoda* whole, although without aids to capture such as the toxicysts of other haptorids. Some support for the raised lip is provided by sheets of microtubules, although these apparently take no part in ingestion.

Other predators have similar capture mechanisms. The rostral process armed with toxicysts which *Dileptus* uses to ingest *Colpidium* (Miller, 1968) and the mobile and deformable anterior end of *Lionotus* are two such cases.

The ability to ingest large prey results in a different feeding pattern which is considerably more erratic than that in small particle-feeders. One prey organism takes a long time to ingest and a large amount of vacuole membrane to enclose it. *Bursaria* is exceptional in that it only takes seven seconds to form a vacuole round an ingested *Colpidium*, but this is too slow to keep pace with the rate at which prey is being drawn into its large buccal cavity in a dense culture. In such a situation, where *Bursaria* cannot eat fast enough, its oral cavity clogs up and much of the prey escapes (Fenchel, 1980b).

Some ciliates develop carnivorous tendencies only when their preferred food supply is exhausted. In the absence of bacteria *Blepharisma* eats its own kind and develops giant forms for self-protection (Giese, 1973). The special capture methods of the suctorian ciliates are sufficiently different to justify a separate section.

Suctorian Ciliates

The suctorians are now firmly established in the Ciliophora in spite of the absence of external cilia in the adult form. They reproduce by budding a dispersive young phase as a simple ciliated organism with an

Figure 8.5: A Suctorian Ingesting its Prey Ciliate. (a) The ciliate is immobilized and held by the tentacle; (b) haptocysts (ha) orientate round the periphery of the tentacle knob and discharge cytolyzing substances into the prey; (c) the pellicles of the predator and prey fuse, the cylinder of microtubules moves up the enlarging tentacle and splays outwards whilst the plasma membrane invaginates to form a feeding tube. =====, cylinder of microtubules.

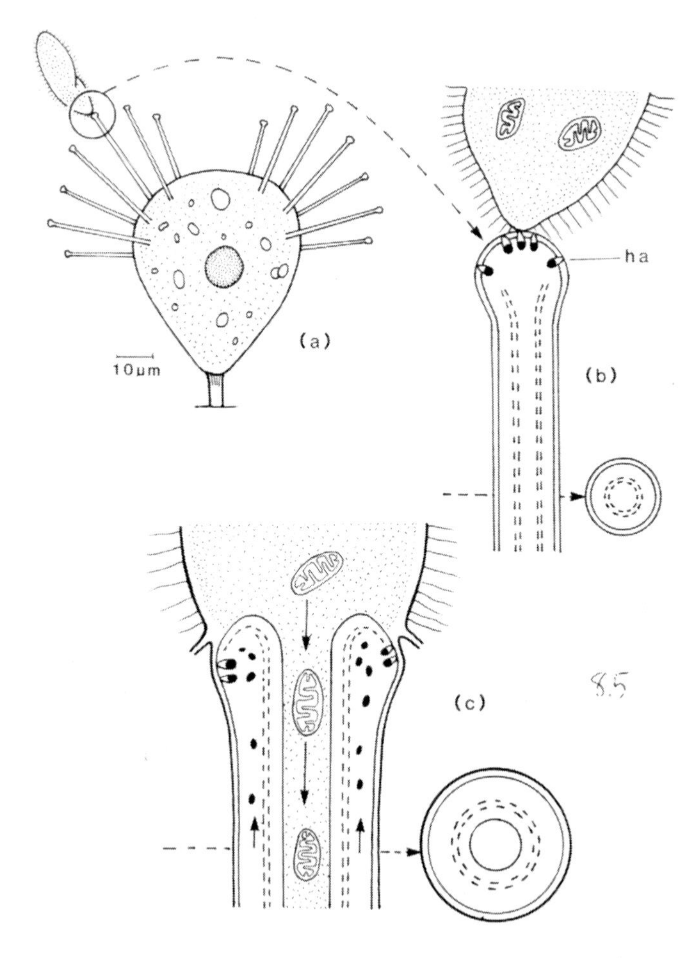

Source: Adapted from Bardele (1972, 1974), Rudzinska (1965) and Sleigh (1973), by permission.

infraciliature resembling other kinetofragminophoreans. On metamorphosis to a sedentary, tentacle-bearing adult, regenerative kinetofragments of barren kinetosomes are retained to supply cilia at the next budding phase. Further evidence of their ciliate relationships is shown in the structure of the tentacles, by microtubule arrangements and the possession of haptocysts, a form of toxicyst (Figure 8.5).

Many suctorians are ectocommensals on a wide variety of marine and freshwater hosts. They use the motile activities of their host or its feeding strategy to bring food to them. In many cases the host is also sessile, perhaps a colonial cnidarian, where the co-operative action of the feeding polyps creates enough disturbance in the surrounding water to attract potential food acceptable also to the suctorian.

Most suctorians are carnivores, although selectively so, for certain live amoebae, flagellates and heliozoans are not acceptable (Hull, 1961a, b). They have developed a special method of feeding in which the tips of the tentacles act as cytostomes (Bardele, 1974). The stimulus for ingestion involves contact between the prey and the tentacles of the predator, followed by a chain of events resulting in the cytoplasm of the prey organism appearing in food vacuoles in the suctorian.

Tentacles as Feeding Organelles

The internal organization of the suctorian tentacle and its mode of action have aroused much interest as a result of several ultrastructural studies. The arrangement is functionally of one tube inside another, separated by a characteristic array of microtubules (Bardele, 1974; Hitchin and Butler, 1973, 1974; Rudzinska, 1965, Tucker, 1974). The idea that the microtubules in the tentacle might have some other function than simply support led to theories of how prey cytoplasm could be conducted down what was virtually an extended feeding tube (a tentacle) and into the body of the predator. Although the majority of suctorians have tentacles which are basically similar and function both for prey capture and ingestion, a small group of marine suctorians, of which *Ephelota gemmipara* is an example, show specialization into feeding and prehensile tentacles. Descriptions of the structure and mode of function of tentacles in *Tokophrya infusionum*, *Acineta tuberosa*, *Podophrya collini*, *Dendrocometes paradoxus*, *Discophrya*, *Heliophrya erhardi* and *Choanophrya infundibulifera* are all available in the literature (Bardele and Grell, 1967; Curry and Butler, 1979; Hauser and Van Eys, 1976; Hitchin and Butler, 1973;

Hull, 1961; Rudzinska, 1970, 1974).

Most suctorians respond only to random collision by living ciliates, and no feeding response is evoked by ciliates killed by various physical means such as heat, freezing and thawing, or increased hydrostatic pressure. This is a more precise response than in many other protozoa that feed by phagocytosis of inert solid food. Small ciliates such as *Tetrahymena* are suitable prey and have been used extensively in feeding experiments on *Tokophrya* (MacKeen and Mitchell, 1977). *Tokophrya* has two clusters of tentacles, very slender and straight, about 20-30 μm long, each ending in an inflated knob which is covered by a single membrane (the plasma membrane).

When suitable prey makes accidental contact, organelles of attack (haptocysts or missile bodies) stream up the outer tube and arrange themselves around the periphery of the knob, projecting through the plasma membrane and discharging a battery of suctorian enzymes. The haptocysts, although recognized as a form of toxicyst, are unique to the group; their presence has been confirmed in *Acineta*, *Ephelota*, *Dendrocometes* and *Heliophrya*. Structures similar to haptocysts in *Cyathodinium* have been used as evidence that this ciliate should be classified with the Suctoria (Paulin and Corliss, 1969). Indirect evidence for the attacking function of haptocysts is that they are absent from the detritus feeder *Choanophrya*, which does not need to subdue its prey (Hitchin and Butler, 1973). This organism, a commensal on the mouthparts of crustaceans, catches food by expanding its tentacles into funnels which pivot backwards and forwards. These feeding tentacles then develop an invaginated membrane tube as do other suctoria (Hitchin and Butler, 1973).

The battery of enzymes secreted by haptocysts facilitates penetration of the predator tentacle into the body of the prey by dissolving a small area of pellicle. At the same time this immobilizes the prey by blocking ciliary action and cytolizing the prey cytoplasm in the immediate vicinity of the tentacle down which it can be drawn into the body of the predator. One or more tentacles may be involved in feeding, and the length of time taken to ingest the contents of the prey organism is dependent on the number of tentacles involved.

The ingestion process is preceded by a change in the structure of the tentacle knob (Figure 8.5). The pellicles of the predator and prey fuse to seal the edges of the puncture site, and the cylinder of microtubules moves upwards and splays out into the knob, making a funnel to enlarge the collecting area. The surface membrane starts to invaginate down the tentacle, forming a feeding tube internal to the microtubule ring. Partially-predigested fragments of pellicle membrane, cilia and mitochondria move down with the invaginating membrane deep into the predator where an endocytotic vacuole begins to form. These vacuoles are pinched off in the same way as in any ciliate phagocytosis, with the internal tube of the tentacle acting as an extended cytopharynx. Having established an inward flow of fluidized, partially-digested material, the succeeding endocytotic vacuoles are seen to contain intact mitochondria and other prey organelles drawn from beyond the influence of the initial haptocyst enzyme secretions.

Flow of particles down the tube continues at a constant and uniform rate in all the tentacles so involved and continues until the prey has been emptied. The rate of flow of the prey cytoplasm is temperature-dependent, the rate in *Podophrya* being 1 μm per second at 18°C and in *Choanophrya*, 10 to 20 μm per second at 18°C for smaller particles and slower with larger particles. The temporary nature of the ingestion tube should be emphasized; the lining membrane disappears and the tentacle reverts to normal after feeding is completed.

From the instant the plasma membrane starts to invaginate down the tentacle, it is being used up in producing endocytotic vacuoles internally. The need for a sudden and rapid production of membrane material to line the long passage as feeding proceeds is similar to the situation in herbivorous ciliates whilst ingesting algal filaments. Various vesicles have been implicated in membrane production; the membrane-bound osmiophilic vesicles with a striated matrix which move up the outer tube at the start of ingestion are the most likely candidates (Rudzinska, 1970). A mechanism for moving the lining membrane and ingested food down the interior of the tentacle must exist. Evidence comes from the microtubule pattern in the tentacles: two concentric rings of microtubules which describe a steep helical path: individual tubules are held in position with intertubule links to prevent longitudinal distortion. Microtubule arms project centripetally to make contact with the invaginating plasma membrane and by pivotting from the fixed microtubules, move it down. To achieve this it is necessary to assume that the downstroke of the arm is the propulsive force, similar to ciliary action. Evidence for this type of suctorian movement has been put forward by Bardele (1974), Hitchin and Butler (1973) and Tucker (1974). It is in accordance with descriptive mechanisms of feeding organelles in other ciliates, *Nassula* and *Phascolodon*, where microtubule arms are involved in drawing in food.

Endocytotic Vacuoles and Digestion

The process of digestion in endocytotic vacuoles in suctorians does not differ markedly from that in other ciliates that feed more conventionally using a single cytostome. However, suctorians have no mechanism for extruding spent vacuoles, which remain as residual vacuoles indefinitely and may be the cause of ageing in the organism (Rudzinska, 1970).

Digestive vacuoles are pinched off from the inner end of the tentacle, starting with those containing rather fluid, predigested prey organelles. The sequential size changes in digestive vacuoles are consistent with the general description in Chapter 6, with a maximum size of 5 μm being reached after 4 hours and a gradual shrinkage over 24 hours to residual vacuoles of 2 μm diameter. The early appearance of acid phosphatase in the ingestion tube and in newly formed digestion vacuoles suggests that the initial enzyme is actually ingested from the prey and not produced by the predator suctorian (Rudzinska, 1972).

As with other protozoa, suctoria are capable of autodigestion when starved, forming autophagic vesicles which contain quantities of acid phosphatase (Rudzinska, 1973).

The intrinsic interest of the suctorians lies much more in their methods of trapping food than in the succeeding digestive processes.

Ciliates in Ruminants

Ciliates form a large proportion, some 40 per cent, of the biomass of micro-organisms that inhabit the rumens of herbivorous bovids and cervids and many other herbivorous mammals. Because they are invariably present in large numbers (10^6 ml^{-1}), and in healthy animals, it is assumed that they do nothing but good. It has proved difficult to quantify the contribution made by protozoa to the host's energy budget because in their absence rumen bacteria compensate by increasing in numbers. Recent reviews of the metabolic activities of this interesting group discuss the role of protozoa in the ruminant (Bauchop, 1980; Clarke, 1977; Coleman, 1975, 1979; Hungate, 1966, 1978). The intricate pattern of anaerobic syntheses and breakdown processes involving the whole micro-organism population (protozoa and bacteria) is a unique one.

Ciliates in the rumen divide into two different orders, Trichostomatida and Entodiniomorphida, on morphological grounds, a division which is also supported by biochemical and physiological evidence (Figure 8.6). Typical of most ciliates they possess a cytostome and food-trapping mechanisms, and some form of heterotrophic nutrition.

Figure 8.6: Ciliates in the Ruminant. (a) Trichostome ciliates, *Isotricha* and *Dasytricha*; (b) amylopectin reserve in granular form; (c) and (d) entodiniomorphids, (c) *Polyplastron* showing two of the skeletal plates and a large macronucleus, (d) *Entodinium*, several species stained to show the large macronucleus.

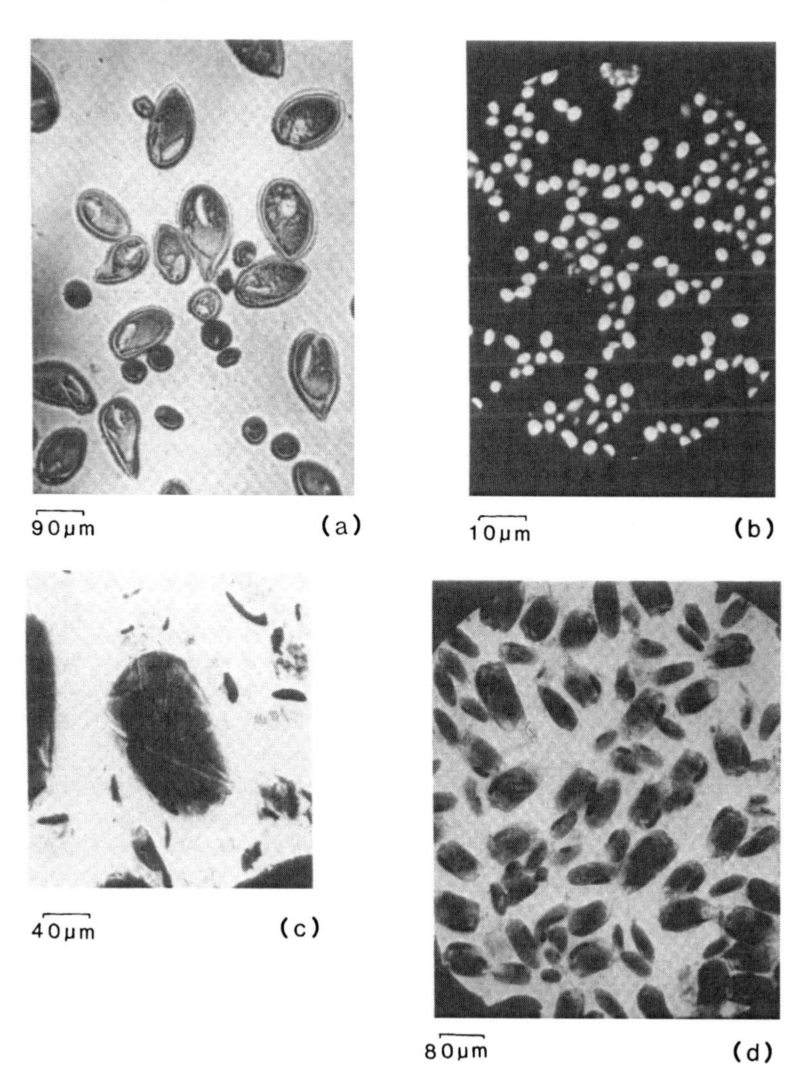

They utilize a range of naturally-occurring substrates from plant material, cellulose, hemicelluloses, fructosans, pectin, starch, soluble sugars and lipids and actively ingest rumen bacteria (Coleman and Laurie, 1974).

Numbers and types of ciliates vary with the diet of the host. The largest populations occur in domestic ruminants as a result of their highly digestible diets and the most restricted populations are found in wild browsing animals. This population analysis can be carried further: trichostomes increase when the diet has increased soluble carbohydrate; *Entodinium* species increase on a starch-containing diet. However, types of populations are not controlled by diet alone. Populations of the larger entodiniomorphs stabilize into two distinct types which are mutually exclusive (Eadie, 1967). Type A consists of *Polyplastron*, *Diploplastron* and *Ophryoscolex* and Type B of *Eudiplodinium*, *Epidinium*, *Eremoplastron* and *Ostracodinium*. The predatory activities of *Polyplastron* destroy all Type B species except certain robust individuals.

Trichostome Ciliates

These are represented by the genera *Isotricha* and *Dasytricha* which have uniform ciliation interrupted only by the position of the cytostome (Figure 8.6(a)). Although pinocytotic uptake of nutrients is clearly an important activity, little information is available on the location of pinocytotic pits.

Isotricha possesses an attachment organelle anteriorly on its dorsilateral surface (Orpin and Letcher, 1978). Such an organelle, however ill-defined, serves to attach the protozoa to fragments of food material, especially fragments which contain soluble sugars. Sugars diffusing from the food elicit a chemotactic response in *Isotricha* providing the sugar concentration reaches a certain critical threshold and that soluble plant protein at a concentration greater than 20 $\mu g\ ml^{-1}$ is present in the medium. The ability to attach to food particles and to the rumen walls is considered to be of ecological advantage to the ciliates in preventing them from being carried out of the rumen too readily in the digestive flow.

The nutritional requirements of these ciliates have been investigated mainly in *in vitro* experiments. There are some differences in the carbohydrate metabolism and enzyme systems of *Dasytricha* and *Isotricha* that may be related to differences in size of the two genera. The larger *Isotricha* ingests particulate food such as plant starch grains (or rice starch experimentally). With its very small cytostome *Dasytricha* rarely ingests starch grains, although there is no reason to believe that the appropriate enzyme systems could not be developed. Both genera engulf bacteria from the rumen and in culture *in vitro*, although their choice might be size-dependent (Sugden and Oxford, 1952; Wallis and Coleman, 1967).

Table 8.1: A Comparison of Carbohydrate Utilization in *Dasytricha* and *Isotricha*.

Substrate	*Dasytricha*	*Isotricha*
Plant starch grains	–	+
Glucose	+	+
Fructose	+	+
Sucrose	+	+
Cellobiose	+	–
Salicin	+	–
Galactose	+	–
Mannose	+	–
Maltose	+	–
β-glucosides	+	+ (slowly)
Raffinose	+	+
Inulin	+	+
Bacterial laevan	+	+
Melibiose	+ (slowly)	+ (slowly)

Note: The results are of *in vitro* tests of carbohydrate utilization by either intact organisms or cell free extracts. +, utilized to moderate or full extent; –, not utilized.

Trichostome ciliates are admirably adapted to a diet of plant material, being furnished with a whole range of enzymes which allows them to utilize the soluble carbohydrates (and starch grains) available in the rumen after feeding (Table 8.1). Both genera can utilize saccharides containing one or more hexoses; therefore glucose, fructose and sucrose appear on the list of acceptable substrates. The speed at which sugars are taken up varies considerably. Placed in order glucose and fructose top the list with sucrose next. *Dasytricha* is perhaps the most interesting of the two genera in that it can metabolize the plant sugars, cellobiose and salicin, and a β-glucoside available to it when the diet is a poor quality fodder and low in glucose and fructose. *Dasytricha* seems to have enzyme systems more adapted to extreme conditions in the rumen than does *Isotricha*.

This group of ciliates is unique in its ability to utilize available carbohydrate at an extremely rapid rate, whether the process is one of conversion to storage polysaccharide or is an energy-producing fermentation. For instance, starved organisms will convert a large proportion of extracellular glucose to amylopectin granules for temporary storage within five minutes of presenting this substrate. The granules, 2 by 3 μm in size, apparently fill the organisms, rendering them swollen and heavy. Neither genus appears to have a mechanism for limiting the uptake of glucose when surplus is provided experimentally, for they continue to synthesize amylopectin granules until they rupture. Under more normal feeding conditions *Isotricha* may have up to 70 per cent of its dry weight as amylopectin reserve. Figure 8.6(b) shows the appearance of amylopectin granules released from rumen ciliates. Rather unexpectedly a dried preparation of amylopectin extracted from these ciliates is not utilized when fed back to the same group of species (Howard, 1959).

Entodiniomorphid Ciliates

Entodiniomorphid ciliates include a number of small and large, structurally more complex rumen ciliates (Figure 8.6(c, d)). They have no somatic ciliature in kineties. The cilia are restricted to one or two zones, the circumoral or vestibular cilia enclosing the cytostome and the aboral ciliary tuft (if present) at some other location. The vestibular cilia perform a dual role, functioning as a feeding structure and as a locomotory organelle, while the aboral cilia are solely an accessory locomotory structure for large and cumbersome entodiniomorphids. They are absent in the smaller genera. The need to move in search of food is not of prime importance to the rumen ciliates as food is brought to them by the ruminating activities of the host animal in the form of thick soupy liquor containing plant fragments, starches and sugars.

A large number of species of entodiniomorphid ciliates have been described and more are being recorded as different herbivores are being examined. One of the more recent species lists published is that from the Brazilian Water Buffalo by Dehority (1979) in which he recorded 49 different species of ciliates. These include eight new species, three belonging to the genus *Entodinium*, four to the genus *Diplodinium* (*Ostracodinium*) and one to *D. (Eudiplodinium)*. However, much of the experimental work has been restricted to a small number of species which have been successfully cultured *in vitro*, particularly *Entodinium caudatum* (Coleman, 1969), *Epidinium ecaudatum caudatum* (Coleman and Laurie, 1974, 1976; Coleman *et al.*, 1972, 1976) and *Polyplastron multivesiculatum* (Bonhomme *et al.*, 1982; Coleman *et al.*, 1972).

The cytostome of entodiniomorphids is normally anterior and terminal, leading to a depression, the cytopharynx, with a system of fibrils for conducting endocytotic vacuoles into the interior of the cell. One feature of entodiniomorphids that is probably unique to this group is the visible demarcation of zones within the cytoplasm. The central area of cytoplasm is a digestive area or gastric sac into which ingested food is transported for initial breakdown and from which indigestible waste is discharged through a permanent cytopyge. Storage of synthesized amylopectin granules is restricted to the peripheral layer of cytoplasm and within the framework of the skeletal plates, conspicuous structures lying beneath the pellicle in the larger entodiniomorphids (Figure 8.6(c)). The use of the term skeletal plate implies a supportive function, which is hardly necessary in organisms that have a tough and non-deformable pellicle.

Entodiniomorphids show some food preferences. The smaller *Entodinium* species preferentially ingest starch grains. Only the larger entodiniomorphids such as *Epidinium*, *Polyplastron* and *Diplodinium* are seen to contain fragments of plant material. The predatory activities of *Polyplastron* do not provide the necessary nutrients to prolong life (Eadie, 1967).

Entodinium caudatum utilizes plant starch grains and also maltose and glucose, as raw materials for synthesizing storage polysaccharide up to 6-7 per cent of the dry weight of the organism (Sugden, 1953; Abou Akkadar and Howard, 1960; Coleman, 1969). Particulate food, starch grains and bacteria are engulfed through the cytostome and pass into the gastric sac, but as in other protozoa soluble sugars enter directly through the plasma membrane. Whether this process is active or passive depends upon the concentration of glucose in the external medium. Glucose in low external concentrations is predominantly taken up by an active process into the endoplasm and used as a raw material for polysaccharide synthesis, but when the external concentration exceeds 2.5mM diffusion is predominantly passive. By whatever route glucose enters *Entodinium* it proceeds along the following series of stages: glucose → free glucose in the cytoplasm → maltose → hexose monophosphate → amylopectin. The pattern of carbohydrate utilization in *Entodinium* differs from that of *Isotricha*, in that *Entodinium* has an absolute requirement for granular starch expressed in terms of prolonging the life of the protozoa.

The larger entodiniomorphids utilize plant fragments (and experimentally, powdered cellulose) as a raw material for amylopectin synthesis, but it is still not confirmed whether cellulases for the initial degradation are produced by the protozoa or by their intracellular bacteria (Sugden, 1953; Hungate, 1942, 1943). Coleman and coworkers isolated and grew six species of entodiniomorphid ciliates with powdered dried grass as the sole carbohydrate source. However they recognized that this neither proves that the protozoa were relying entirely on cellulose nor that they were producing their own cellulases (Coleman *et al.*, 1976).

Epidinium has been shown, however, to possess α-galactosidase, an enzyme necessary for the hydrolysis of chloroplast constituents. This enzyme is absent from the non-grass-eating trichostome ciliates (Bailey and Howard, 1963). Thus there may be some enzyme adaptation in *Epidinium*. Further support for possible cellulase activity comes from the finding that hemicellulases have been found in cell-free extracts of *Epidinium ecaudatum* (Bailey and Gaillard, 1965).

Carbohydrate Metabolism

Rumen ciliates contain a wide range of carbohydrases, which allows them to utilize a variety of substrates. Table 8.2 lists the enzymes identified in the three kinds of rumen ciliates, the trichostomes and small and large entodiniomorphids.

The conversion of glucose, maltose and, indirectly, soluble starch, to amylopectin by *Epidinium ecaudatum* follows a normal phosphorylation pathway mediated by hexokinase, which can be demonstrated in cell-free extracts of the organism (Coleman and Laurie, 1976). A similar phosphorylation pathway also occurs in *Isotricha*, *Dasytricha* and *Entodinium caudatum* (Oxford, 1955; Coleman, 1969). This phosphorylation pathway may be illustrated very simply by the following scheme:

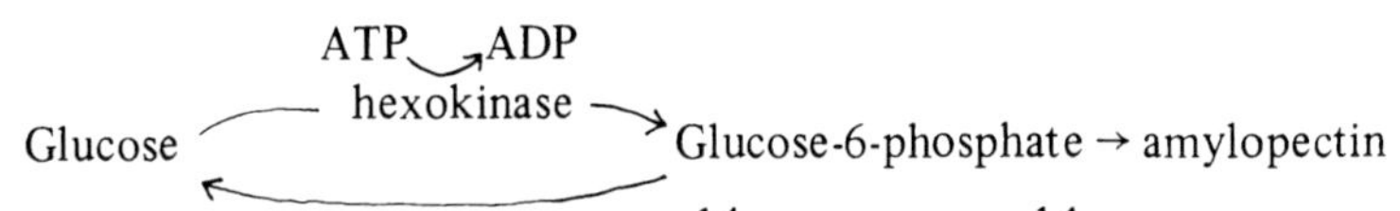

When *Epidinium* is incubated with ^{14}C glucose the ^{14}C label appears in three places, in the cytoplasm as free glucose, in amylopectin by biosynthesis as above and in intracellular bacteria which are liberated by sonification during preparation of the cell fractions. To explain this it is postulated that glucose is taken into the cytoplasmic pool initially unchanged and then is rapidly converted to glucose-6-phosphate. This may be subsequently hydrolyzed back to glucose, a major constituent

Table 8.2: Enzymes Shown to be Present in Rumen Ciliate Protozoa.

Enzyme	*Dasytricha Isotricha*	*Entodinium*	*Epidinium*
α-amylase	+	+	+
Maltase	*Dasytricha* only	+	+
Levanase	+	–	–
Invertase	+	+	–
Inulinase	+	–	–
Cellobiase	*Dasytricha* only	–	–
β-glucosidase	*Dasytricha* only	–	–
Maltose phosphorylase	?	?	+
Hexokinase	+	+	+
Phosphoglucomutase	?	?	+
Glucose-6-phosphatase	?	?	+
Pectin esterase	*Isotricha* mainly	–	–
Polygalacturonase	+	?	?
α-galactosidase	–	–	+
Isomaltase	–	–	+
Cellulase	–	–	?

Note: The results are of *in vitro* experiments using cell-free extracts of rumen protozoa. +, presence of the enzyme is confirmed; –, enzyme found to be absent; ?, no information available.

of the second sugar pool. Most of the free sugars and their derivatives are found in this pool where they are intermediates in the synthesis or degradation of polysaccharides. Maltose and soluble starch follow similar routes, but appear to be less effective as a source of carbohydrate.

The overall picture of carbohydrate utilization in rumen ciliates is remarkably constant. The pattern is one in which a proportion of the carbohydrate taken in is converted to amylopectin for temporary storage and the remainder is anaerobically and directly fermented in an energy-producing process into lactic, butyric and acetic acids (normally in equal proportions) with the production of a quantity of carbon dioxide and hydrogen (Gutierrez, 1955: Howard, 1959; Abou Akkadar and Howard, 1960). These volatile fatty acids diffuse out of the rumen into the bloodstream of the herbivore and provide a source of carbohydrate.

Protein Metabolism

Rumen ciliates obtain most of their amino acids for protein synthesis by ingesting living or dead bacteria, although they can also take up some amino acids directly. The live bacteria may be maintained by entodiniomorphids as temporary populations of intracellular bacteria to form a gradual and continuing source of protein constituents.

Some ciliates are selective in their choice of bacteria and in their treatment of those ingested. *Entodinium caudatum* will engulf any bacteria; *E. simplex* takes *Klebsiella aerogenes* and *Proteus mirabilis* selectively, using their amino acids unchanged for incorporation into protozoal protein (Coleman, 1972). The live bacteria are taken into vesicles for digestion, but how long they survive depends on how quickly they can synthesize a protective capsule of glucose-containing polysaccharide. They may still be viable after 22 hours. *Epidinium*, on the other hand, lyses bacteria in the medium preferentially and ingests the components (Coleman and Laurie, 1974).

The net result of the ciliate-bacteria interaction is that bacterial protein is broken down to amino acids which are recombined to protozoal protein for growth. Any surplus amino acids are released into the medium and are used by free bacteria for their growth.

Rumen ciliates can hydrolyze casein and glycine into peptides and amino acids but cannot then use these for synthesizing protozoal protein except through the agency of intracellular bacteria. Dependence on bacterial synthetic activities seems inescapable.

Lipid Metabolism

Ciliates may be making a significant contribution towards the breakdown of lipids in the rumen, but evidence is erratic. The possible involvement of bacteria has not been convincingly eliminated (Coleman, 1979). Hydrolysis of triglycerides and the hydrogenation of long-chain fatty acids do occur in the rumen and some of this activity can be attributed to entodiniomorphid and trichostome ciliates.

Exocellular Enzyme Secretion

Isotricha and *Dasytricha*, with their demonstrated reliance on soluble nutrients, have developed ways of maximizing supplies of solutes. When the nutrient level of the host animal's fodder is low, particularly with respect to soluble sugars, the ciliates actively secrete exocellular enzymes to break down the more resistant polymers (Thomas, 1960). *Dasytricha* and, to a lesser extent, *Isotricha*, secrete exocellular invertase into their immediate environment, thus ensuring a supply of glucose and fructose which the ciliates need for synthesizing storage amylopectin. In addition *Isotricha* releases pectin esterase, polygalacturonase and several glycosidases and *Dasytricha* releases β-glucosidase (Williams, 1979).

Exocellular enzyme secretion is not restricted to *Dasytricha* and *Isotricha*. The ability of *Epidinium* to lyse bacteria before ingesting them has already been mentioned (Coleman and Laurie, 1974). This demonstrates that *Epidinium* can secrete exocellular enzymes and there is no reason to believe that other entodiniomorphids cannot do likewise.

Micro-organisms and the Ruminant

Evaluating the respective roles of protozoa and bacteria in the ruminant is an ongoing discussion topic. The ruminant with ciliates appears to grow better than a defaunated animal, yet the facts are difficult to explain. In the absence of ciliates bacteria increase in numbers, for they have less competition for food and they are not being predated upon. This increase compensates in part for the absence of protozoa.

The contribution made by protozoa to the host nitrogen economy is not as great as one might expect from the numbers present in the rumen. Only about 2 per cent of the nitrogen in roughage consumed by the host actually passes into its duodenum as protozoal nitrogen, perhaps because of the ability of the ciliates to resist being washed through.

Beneficial effects of the ciliates in a ruminant are probably indirect ones, operating on the bacterial population initially. (1) Their capacity to produce amylopectin reserves rapidly may be important when the host is consuming low-quality fodder, as a means of providing a steady flow of substrates for the microbial population. When food is plentiful, conserving substrate would be irrelevant. (2) They promote an actively-growing population of bacteria by their predation. Actively-growing bacteria maintain reducing conditions in the rumen environment, which are essential for protozoan growth.

Endosymbiotic Relationships

The possible involvement of endosymbionts in eukaryote evolution has been discussed in Chapter 5 in the light of Margulis' (1981) theory. There is no doubt that a close relationship develops between extant heterotrophic species and their autotrophic green algae. The sort of question one could ask is why does *Paramecium bursaria* not normally digest the unicellular alga, *Chlorella*, which occupies vacuoles in its cytoplasm. The nature of its immunity from attack by host lysosomes at certain times, yet not all the time, is difficult to comprehend. There are various ways in which an enclosed organism can protect itself from attack.

It seems that once *Chlorella* is established in a vacuole, the vacuole membrane is altered in some way to repel lysosomal attack (Karakashian and Rudzinska, 1981). However, during the acquisition of a population of symbionts an alternative food source must be available otherwise all the algae are likely to be digested. Only about 20 per cent of the algae do survive to establish a symbiotic relationship (Karakashian, 1975). When the initial stage is passed the symbionts feed and reproduce to keep pace with the rate of division of their host. The algal symbionts of *Climacostomum virens* are not apparently completely enclosed in vacuole membrane, yet they seem to be immune from host lysis (Peck *et al*., 1975).

Endosymbioses between ciliates and bacteria probably occur more frequently than is currently appreciated. *Paramecium aurelia* houses bacteria which it inherits from the parent organism at division, although no clear benefit to the host has been demonstrated (Preer, 1975), the exception being perhaps the killer strain of *P. aurelia* where its endosymbionts (kappa particles) kill competitive strains.

A strong case has been made for an exchange of metabolic activities between ciliates such as *Parablepharisma* which live in anaerobic, high hydrogen sulphide environments and their endosymbiotic bacteria (Fenchel *et al*., 1977). Fermentative pathways of the ciliates produce end-products (organic acids, hydrogen and carbon dioxide), which are utilized by the bacteria, and a similar feedback must occur from the bacteria.

The hypotrich ciliate *Euplotes aediculatus* contains bacteria-like particles which are essential for growth and division. There are 900-1,000 particles, which are destroyed by penicillin treatment, the resulting loss preventing division in *Euplotes*. One can speculate that destruction of the symbionts removes an essential contribution to the metabolism of the host.

In Summary

Some organelles in filter-feeding, carnivorous and herbivorous ciliates have been described without attempting a comprehensive review. The special relationship between the ciliates of ruminants and their hosts and the range of polysaccharides utilized by these ciliates form a large section. While recognizing that suctorians with their tentacles specially adapted for ingesting particulate material from prey organisms are atypical, they have been included because of their intrinsic interest.

It may seem that this chapter has been too selective in its approach to ciliate feeding. When faced with the variety of structures and strategies, from the simple to the very complex, comprehensive coverage is impossible. Yet selecting the groups about which much has been published is only compounding the situation lamented by Corliss (1979), that of 'the incredible unevenness of our knowledge about the forms which have already been described'.

9 VARIETY IN THE MASTIGOPHORA

Mastigophorans, or flagellated protozoa (the 'whip-bearers'), probably show a wider range of feeding and nutritional types than any other group of protozoa. Taxonomically they are included with the Opalinata and the Sarcodina in one phylum Sarcomastigophora, containing organisms which have a single type of nucleus and which have locomotor organelles, pseudopodia or flagella, or both. Sexual reproduction, if it occurs, involves the production of gametes and conventional gametic fusion. This set of criteria indicates a phyletic relationship, but other criteria easily separate the Mastigophora, Opalinata and Sarcodina into distinct subphyla.

The vast majority of flagellates are much less familiar organisms than are the ciliates. It is unfortunately tempting, especially in wide-spectrum ecological investigations of microfauna, to identify individual ciliate species and group all the flagellates together as small colourless flagellates. Small size and comparatively simple external morphology perpetuate this attitude and make identification keys difficult to construct and use. However, this subphylum also contains many organisms which are quite distinctive in shape and colour, and in the possession of complex flagellated organelles. Others are pathogens of considerable economic importance, resulting in the Kinetoplastida in particular attracting a great deal of attention.

The main aim of this chapter is to present and describe the two mastigophoran classes, Phytomastigophorea and Zoomastigophorea, although in a rather selective way, but it will also include a brief mention of the Opalinata. The opalines appear as a rather anomalous group of organisms, difficult to place because they have a superficial resemblance to ciliates.

The Phytomastigophorea are typically chloroplast-bearing photoautotrophs with the capacity to respond to differing light intensities by adjusting their position in the water (see Chapter 4). Nutritional versatility is frequently present. Colourless species do occur, as in the Euglenida, where morphological similarities are very evident and no difficulty is experienced in establishing a relationship between pigmented and non-pigmented species. Ten orders of phytomastigophoreans are currently recognised: Cryptomonadida, Dinoflagellida, Euglenida, Chrysomonadida, Heterochlorida, Chloromonadida, Prymnesiida, Volvocida, Prasinomonadida and Silicoflagellida.

The Zoomastigophorea are obligate heterotrophs, often parasitic, and although by definition they are flagellated, some spend part of their lives in an amoeboid form devoid of externally visible flagella. The orders included are Choanoflagellida, Kinetoplastida, Proteromonadida, Retortamonadida, Diplomonadida, Oxymonadida, Trichomonadida and Hypermastigida. Obligate heterotrophy is found in all zoomastigophoran flagellates; none is pigmented and all practise some form of endocytosis. Many are parasites of body cavities, the alimentary canal, or blood, in which organic substrates are abundant and complete utilization of the substrate is unnecessary. This results in the type of abbreviated metabolic pathway found in the blood form of *Trypanosoma* species (Vickerman, 1965).

Consistent with the wide range of nutritional types found in the Mastigophora, methods of intake of raw materials are equally varied, depending to a great extent on the size of the molecules or particles involved. The presence of a distinct cytostome is not a prerequisite for heterotrophs, and ingestion may take place at a relatively unspecialized point on the body surface. Inorganic or small organic molecules enter the body by diffusion or active transport. In those species that ingest solid food, various methods of food-trapping are employed, involving the use of flagella for creating currents of water for the active filtering of small particles (Figure 5.3).

Phytomastigophorea

The phytomastigophorans are mainly free-living in fresh and salt water, with only a few adopting a parasitic way of life. They usually have only one or two flagella. Most of the ten orders contain species which are autotrophic or heterotrophic or reversibly both, with and without photosynthetic pigments.

Many of these green, red, yellow or brown pigmented autotrophs are as closely allied to the algae as they are to protozoa. The volvocid, *Chlamydomonas moewusii*, is an obligate photoautotroph requiring only inorganic materials and sunlight as an energy source (Hutner and Provasoli, 1955). In others autotrophy has to be supplemented by heterotrophic nutrition with organic substrates. *Ochromonas malhamensis* (Chrysomonadida), possibly due to its low chlorophyll content, relies more on organic compounds as an energy source. Other flagellates prefer to keep their options open and adjust according to substrate availability and energy source. *Euglena gracilis* may switch from autotrophy to heterotrophy in the absence of light energy, provided a suitable carbon source (acetate or ethanol) is available.

Figure 9.1: Carbon Flow in 'Acetate Flagellates' during Dark-induced Heterotrophy. The exogenous substrate may be acetate or a simple fatty acid or alcohol, and the reserve polysaccharide may be paramylon (Euglenida) or starch (Cryptomonadida and Volvocida).

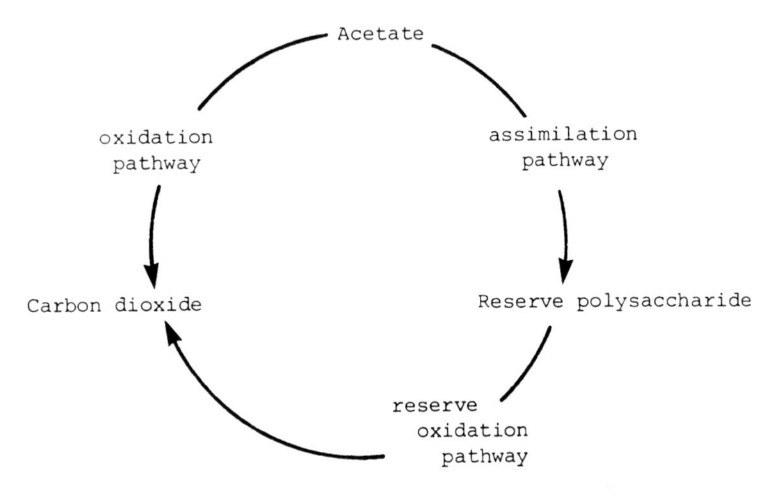

Source: Adapted from Danforth (1968).

The Euglenida, the Cryptomonadida and the Volvocida (Phytomonadida) are frequently grouped together as the acetate flagellates, a description first applied by Pringsheim (1937) on finding similarities in their metabolic abilities. Their preferred carbon sources are acetates, simple fatty acids, and alcohols. Carbohydrates and amino acids do not provide a suitable carbon source for energy or growth in the cryptomonad *Chilomonas paramecium* (Holz, 1954). Many have in common the ability to switch from photoautotrophy (photosynthesis) to heterotrophic oxidative pathways in the absence of light energy. A simple diagram shows the pattern of carbon flow in acetate flagellates during heterotrophy in the dark, with acetate either being converted into reserve polysaccharide or being used directly as an energy source, according to the needs of the organism (Figure 9.1). A recent review of the metabolism of the acetate flagellates has been published elsewhere (Lloyd and Cantor, 1979).

Intracellular polysaccharide reserves in the Phytomastigophorea are variable but rarely are glycogen, as to be expected in organisms which show closer affinities to algae than to animals. Many store polysaccharide as β-1,3 glucans, which have not been recorded in bacteria, green plants, or the metazoa. Relatively large amounts of reserve polysaccharide are synthesized by some flagellate protozoa, a situation comparable to the rumen ciliates described earlier. *Chilomonas paramecium* (a cryptomonad) and *Polytomella caeca* (a volvocid) store up to 20 per cent of their dry weight as starch. Tables 4.1 and 7.1 have summarized some of the characteristics of the phytomastigophoran flagellates.

Euglenida

The most interesting order is that of the euglenoid flagellates, which show a variety of nutritional characteristics and perhaps an evolutionary sequence from the obligate photoautotroph to the carnivore. *Euglena pisciformis* is an obligate photoautotroph which will not grow in the dark, even when supplied with a rich organic medium including acetic acid. Slightly less restrictive, *Euglena gracilis* is a photoautotroph but reversibly a facultative heterotroph. Next in sequence is *Astasia longa*, an obligate heterotroph which grows best on acetate as an exogenous carbon source at pH 7.0 (Hunter and Lee, 1962). From the permanent loss of photosynthetic ability, as occurs in *Astasia,* the next logical step is the development of a means of ingesting particulate food. The carnivorous habit follows, illustrated by *Peranema trichophorum*, with its permanent cytostome and a feeding organelle (the rodorgan) which it uses in phagocytozing a wide variety of food, from yeasts to living *Euglena* (Chen, 1950). Apart from the possession of a feeding apparatus and the absence of photosynthetic pigment, *Peranema* closely resembles *Euglena* in general morphological characteristics (Leedale, 1967).

The nutrition of that ubiquitous and much used genus, *Euglena*, has been described extensively in several monographs, more especially in the three volume series by Buetow (1968, 1982). *Euglena gracilis* is the most favoured organism for enzyme investigations in the Phytomastigophorea. Consequently metabolic pathways for the synthesis and degradation of carbohydrates most frequently refer to this species. By confirming the presence of the appropriate enzymes, it is possible to say that the metabolic pathways operating in *Euglena* are gluconeogenesis, the oxidative pentose phosphate cycle, the tricarboxylic acid cycle (Krebs cycle) and the glyoxylate by-pass. Paramylon reserves are laid down in granules that may vary widely in size and number. Paramylon is generally considered to be the principal energy reserve in *Euglena*, but at high temperatures particularly, wax esters become proportionately more important. *Euglena gracilis* grown at 15°C contains in 10^6 organisms 291.1 μg paramylon and 14.5 μg wax esters, and at 33°C, only 55.5 μg paramylon to 130.0 μg wax esters (Kawabata *et al*., 1982). The intracellular polysaccharide reserves of the different phytomastigophoran groups have been listed in Table 7.1. *Euglena gracilis* is forced to draw extensively on these paramylon reserves during the early part of the stationary growth phase, with a consequent increase in the activity of β-glucosidase, an enzyme involved in paramylon degradation (Baker and Buetow, 1976).

Figure 9.2: The Ingestion Organelle of *Peranema* in Longitudinal Section, Showing the Double Rodorgan (ro) of Close-packed Microtubules, the Cytostome (cy) and an Area of Phagoplasm (ph) (x 7,500).

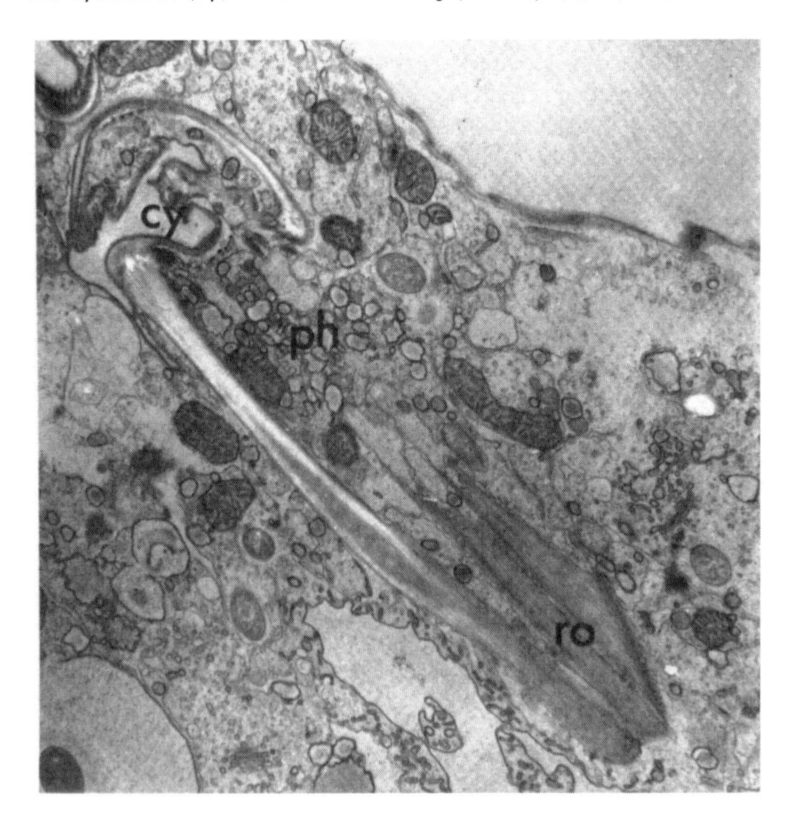

Figure 9.3: A Diagrammatic Reconstruction of the Feeding Apparatus of *Peranema*, in Dorsal View. ca, canal opening; cy, cytostome; cys, cytostomal sac; f_1, anterior flagellum; f_2, trailing flagellum; re, reservoir; CT, canal thickening; DM, double serrated marginal lamella; IA, inner anchoring lamella; IL, inner longitudinal lamella; IR, inner rod; OA, outer anchoring lamella; OL, outer longitudinal lamella; OR, outer rod; SF, striated fibril.

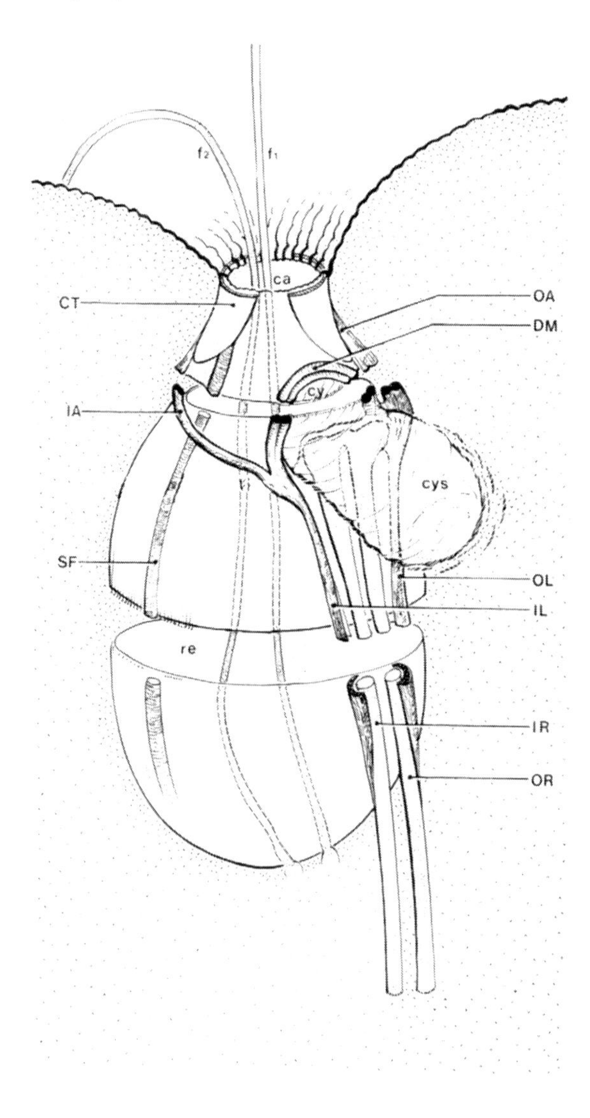

Figure 9.4: ***Peranema***, Showing Helical Striations of the Pellicle and an Extremely Flexible Body.

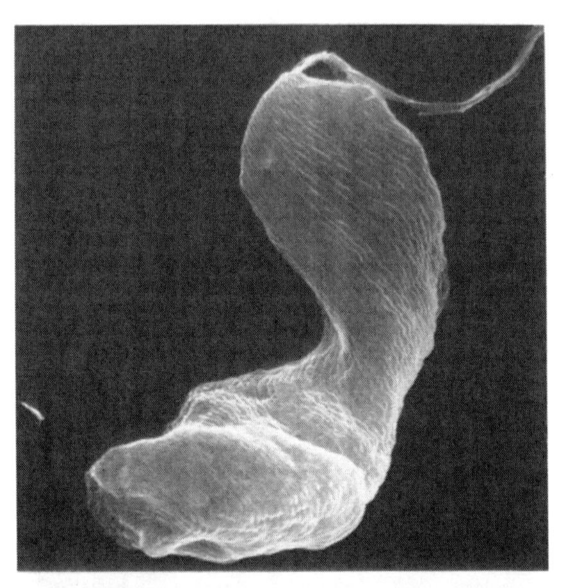

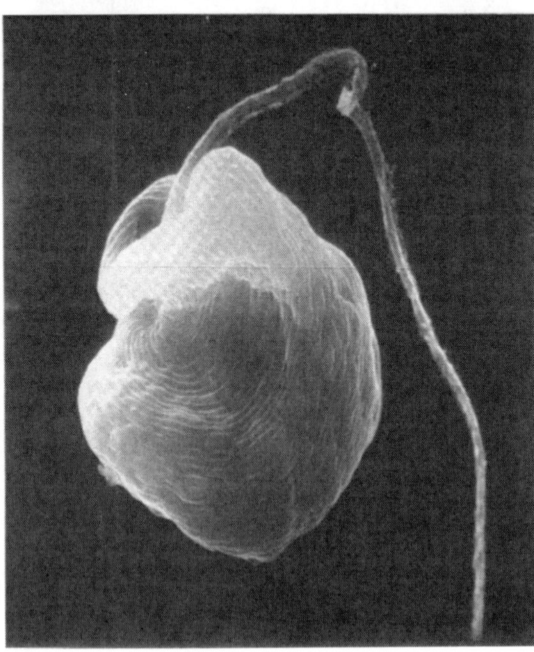

The colourless euglenoid flagellates *Peranema*, *Heteronema* and *Entosiphon* are particulate feeders or carnivores and have developed ingestion organelles associated with a permanent cytostome (Leedale, 1967; Mignot, 1966). The rodorgan of *Peranema* is a double structure that is easily seen in a good light microscope, forming a movable feeding organelle between the cavities of the reservoir and the cytostome (Nisbet, 1974). Each rod contains a core of some 100-200 microtubules embedded in a transparent matrix and enclosed in a sheathing membrane, which unites the pair of rods anteriorly (Figure 9.2). During feeding the rodorgan is moved by an elaborate arrangement of articulating lamellae which is connected to the bases of the pair of rods and to the opening of the anterior canal (Figure 9.3). Direct observation of the ingestion process is made difficult because it is accompanied by violent writhing and twisting of the organism (Figure 9.4). It seems probable that the rodorgan and the cytostome are pulled forwards to fill the external opening of the anterior canal, which expands to allow intake of quite large food fragments. Withdrawal of these structures sucks food into the cytostomal sac, from which endocytosis produces food vacuoles in the standard fashion. The vacuolated phagoplasm, which is associated with an actively-feeding organism, lies between and around the two branches of the rodorgan (Figure 9.2). The ingestion organelle in *Entosiphon* is somewhat similar to that of *Peranema* although consisting of three triangular groups of microtubules enclosing a distinctive fibrillar structure and siphon tube which is responsible for propelling food into the cell (Mignot, 1966; Patterson, 1983; and Figure 9.5).

Providing the required nutrients are present, *Peranema* is not a particularly selective feeder and will ingest bacteria, algae, *Chlorella*, yeast (dried or fresh), starch grains, other live flagellates such as *Euglena*, and suspensions of milk or casein. The nutritional requirements of *Peranema* are more complex than those of the autotrophic euglenoids and include lipid components which have not been fully evaluated.

The majority of colourless euglenoid flagellates are heterotrophs without a distinct cytostomal opening, but which possess the typical euglenoid reservoir cavity lined by a single unit membrane that allows pinocytotic uptake of small molecules from the external environment. Colourless flagellates in general thrive only in a medium with a highly soluble organic content. A sewage outflow is an ideal environment.

Volvocida

The Volvocida, like the Euglenida, normally possess green photo-

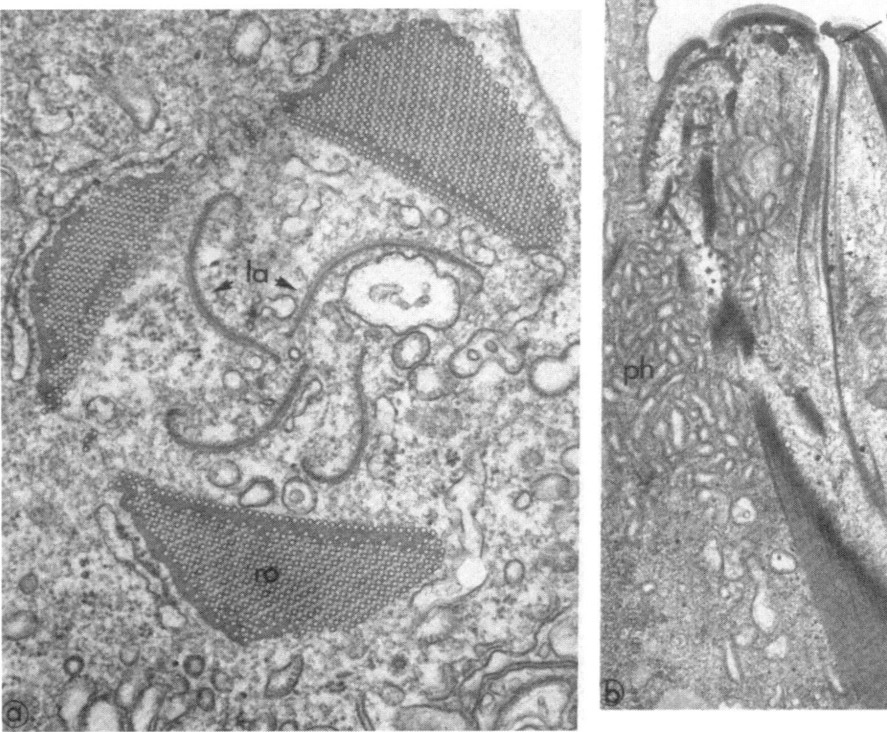

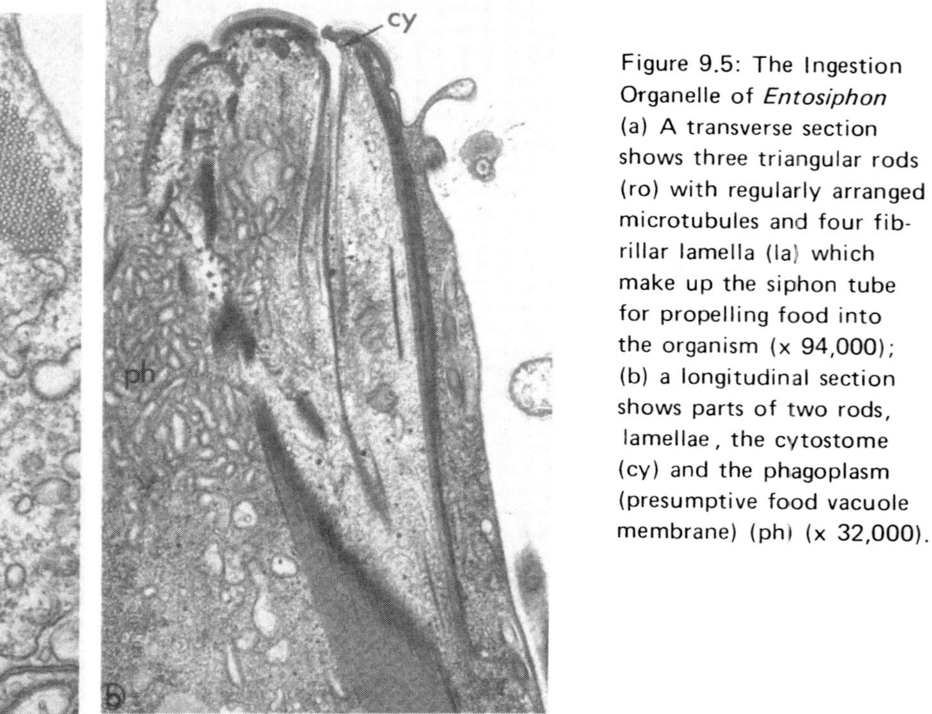

Figure 9.5: The Ingestion Organelle of *Entosiphon* (a) A transverse section shows three triangular rods (ro) with regularly arranged microtubules and four fibrillar lamella (la) which make up the siphon tube for propelling food into the organism (x 94,000); (b) a longitudinal section shows parts of two rods, lamellae, the cytostome (cy) and the phagoplasm (presumptive food vacuole membrane) (ph) (x 32,000).

Source: Electron micrographs by D.J. Patterson, University of Bristol.

synthetic pigments; a few have red haematochrome (*Haematococcus*) as an alternative, and still others are colourless heterotrophs. *Chlamydomonas moewusii* has been quoted as an obligate photoautotroph with a strong light dependence. It will not grow in the dark nor in light intensities less than 15 foot candles (Gross and Jahn, 1962). *Chlamydomonas dysosmos* is more adaptable. It resembles *Euglena* in its capacity for facultative heterotrophy when placed in the dark with acetate as a carbon source. The obligate heterotrophs, such as the unpigmented *Polytomella caeca*, also use acetate as an energy source (Wise, 1959).

The metabolic pathways involved in heterotrophic growth in the volvocids have been described in *Chlamydomonas dysosmos* growing in the dark using acetate as the sole carbon source. These may best be illustrated diagrammatically using a scheme adapted from Neilson *et al.* (1972), exemplifying the utilization of acetate by *Chlamydomonas dysosmos*.

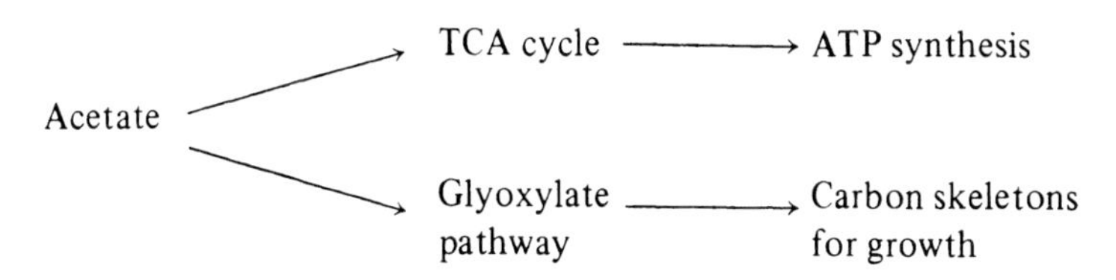

This incorporation of acetate into the tricarboxylic acid cycle is important for energy production. However, equally important is the provision of a carbon source for the biosynthesis of carbohydrates and proteins. A glyoxylate pathway of the type found in *Chlamydomonas dysosmos* is not unique; it figures prominently in biosynthesis in green plants and in the bacterium *Escherichia coli* and in the ciliate *Tetrahymena*, both organisms which can survive on acetate as the sole carbon source.

When *Chlamydomonas dysosmos* makes its metabolic switch in dark culture it must synthesize several adaptive enzymes necessary for successful functioning of the glyoxylate pathway. The two key enzymes in the production of carbohydrate are isocitrate lyase and malate synthetase which require the presence of acetate for their synthesis. The production of these enzymes is blocked during autotrophy in the light.

Cryptomonadida

The Cryptomonadida is a small order of relatively inconspicuous biflagellate organisms, which possess unusual and interesting features.

Although the order includes many chlorophyll-bearing forms, most information is available on the obligate heterotroph, *Chilomonas paramecium*. In common with the other acetate flagellates, *Chilomonas* thrives on acetate as the sole carbon source, but it will also do equally well on succinate. Other acceptable substrates are even-number, straight-chain fatty acids and alcohols (Cosgrove and Swanson, 1952), but starch and related carbohydrates are not utilized (Holz, 1954; Antia, 1980).

A distinctive character of the cryptomonads are the ejectosomes, extrusion organelles which are spaced between the cortical plates (if present) over the surface of the organism, with distinct concentrations in the vestibular region or along the ventral furrow. Functionally they are associated with defence rather than food-trapping, for they occur equally in heterotrophic forms which trap their food and autotrophic forms which do not.

Rhodomonas lacustris is an autotrophic cryptomonad which is well supplied with large ejectosomes. They lie in two or three rows along the ventral furrow (Klaveness, 1981). The photosynthetic pigments in *Rhodomonas*, chlorophylls a and c, carotenoids and phycoerythrin, and reserves of leucosin, indicate a clear affinity with algae.

Chrysomonadida

Chrysomonads are small biflagellate organisms which may be covered with silica scales or some other enclosing envelope. Their tendency to form colonies in which the individuals are held together by a secreted lorica make them important members of the plankton. Many combine heterotrophic and autotrophic nutrition, whilst others are obligate heterotrophs and lack pigment. Heterotrophic members trap food particles in a variety of ways involving flagellar currents aided by pseudopodial ingestion, and often with a protoplasmic collar to concentrate the catch.

Ochromonas malhamensis is an active phagotroph with very limited powers of selectivity, accepting particles which are small and light enough to be drawn in by flagellar lashing. Polystyrene latex particles will stimulate the secretion of acid phosphatase into endocytotic vacuoles as readily as a suspension of bacteria (Dubowsky, 1974). Another unidentified species of *Ochromonas* that behaves similarly, rapidly ingests a suspension of the cyanobacterium *Anacystis nidulans*. This species of *Ochromonas* undergoes a two-stage digestion process. The process begins in the posterior leucosin vacuole and is completed in smaller peripheral vacuoles (Daly *et al.*, 1973). Phagotrophy in

Ochromonas is discussed by Aaronson (1980). In considering the nutritional requirements of *Ochromonas*, its rather limited capacity for autotrophy, and its ability to utilize starch and simple carbohydrates, it seems that this genus is appropriately described as an obligate heterotroph.

Dinoflagellida

The majority of this order are biflagellated members of the plankton, enclosed in elaborately sculptured cellulose envelopes. The three-horned envelope of *Ceratium* is easily recognized in samples of marine plankton (Figure 2.8(d)). If pigmented, then this pigment is yellow or brown and the carbohydrate reserve is starch (Table 7.1) and oil as long-chain unsaturated fatty acids. The accumulation of low-density oil reserves must be regarded as an aid to flotation.

Unquestionably autotrophy plays a very important role in the metabolism of many dinoflagellates, but it is rarely sufficient for their needs. Dinoflagellates have evolved various means of trapping particulate food. This food, often consisting of whole organisms, is engulfed at a specialized oral region in the flagellar (sulcal) groove, between the cellulose thecal plates. It may be drawn in by flagellar currents, trapped by fine cytoplasmic nets, or caught by the action of exploding trichocysts in the colourless *Gyrodinium*. In addition to this form of facultative heterotrophy, there are examples of obligate heterotrophy in many dinoflagellate families.

Noctiluca is an obligate heterotroph with a voracious appetite. It floats as a gelatinous sphere using a long sticky tentacle to trap and convey small crustacean larvae to its permanent mouth near the base of the tentacle (Figure 2.8(e)). *Oxyrrhis marina* is a small active heterotroph which responds to abundant food by building up dense local populations in isolated sea water and brackish pools. It traps bacteria and diatoms in the cavity of the hypocone where they are ingested (Droop, 1954, 1959).

Zoomastigophorea

The existence of obligate heterotrophy in this class of flagellates has already been recognized. A tendency towards a variable (polymorphic) body form, at least in the Kinetoplastida, and often a parasitic way of life, coupled with the absence of chloroplasts, distinguishes zoomastigophorans from the algal flagellates. Although the pattern in some orders is the possession of a single flagellum, many, the Hypermastigida in particular, have numerous flagella associated with an elaborate parabasal apparatus.

Among the free-living zoomastigophorans a fixed body form and a distinct cytostomal area for the ingestion of bacteria are frequently found.

The Choanoflagellida (the collared flagellates) have developed a most efficient food-trapping apparatus. A delicate collar of fine tentacles, enclosed and suspended in a lorica of silica rods or costae, surrounds the base of the single flagellum and acts as a funnel to trap bacteria and detritus drawn in by flagellar undulations (Hibberd, 1975; Laval, 1971; Leadbeater, 1972, 1973; Leadbeater and Manton, 1974). Food material, caught in mucus which covers the ring of tentacles, is then engulfed by advancing pseudopodia. These feeding pseudopodia are usually associated with one or more tentacles and probably receive support from them (Leadbeater and Morton, 1974; Manton *et al.*, 1976). Food ingested by pseudopodia then passes into the body of the organism at some point within the collar, usually between the protoplast and the lorica. Thus the feeding action can be divided into two phases: food-trapping by flagellar currents and collar tentacles, then phagocytosis by pseudopodial engulfment.

Certain zoomastigophoran orders have been selected for inclusion in this section on the basis of medical importance (as parasites) and of biological interest (as symbionts). The Kinetoplastida, Diplomonadida, Trichomonadida and Hypermastigida have been chosen from the group.

Kinetoplastida

The kinetoplastid flagellates have attracted a great deal of attention because they include several genera of considerable economic importance to man and other animals. *Trypanosoma* and *Leishmania*, the causative organisms of sleeping sickness and Kala azar (visceral leishmaniasis) respectively, have been the subject of much research in the fields of ultrastructure, biochemistry and drug resistance. For the most recent review on the biology of the Kinetoplastida reference should be made to Lumsden and Evans (1976, 1979).

The Kinetoplastida includes two sub-orders, Trypanosomatida, which are all parasites of varying importance and pathogenicity, and Bodonida, mostly free-living, ubiquitous and relatively less important. For this reason the bodonids have attracted less research effort.

Figure 9.6: Food-trapping Organelles in Bodonid Flagellates. (a) *Bodo* wafts bacteria into a cytostome (cy) bounded by flap-like lips – the collecting current is produced by a flagellum with a fringe of mastigonemes (ma); (b) *Rhynchomonas* possesses a cytostome at the tip of a mobile scavenging bilobed proboscis (pr) attached to the short anterior flagellum.

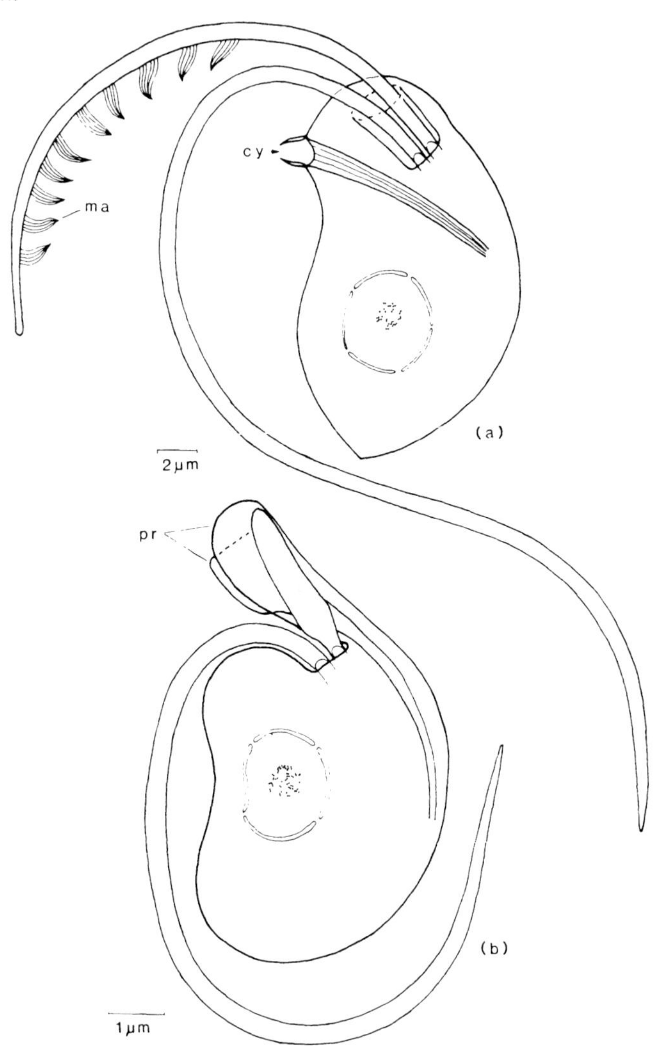

Source: Redrawn from Lumsden and Evans (1976), by permission. Copyright: Academic Press Inc. (London) Ltd.

Bodonids. The bodonids are small and often extremely abundant, especially in bodies of water which have high soluble organic content and bacteria. Food-trapping is achieved in a variety of ways by these scavenging flagellates. Many efficient trapping mechanisms are employed which result in bacteria and detritus being swept into the cytopharyngeal tube. Some *Bodo* species, relatively sedentary organisms, waft bacteria and small particles into a lateral cytostome by movements of the anterior flagellum (Figure 9.6(a)). These movements are made more effective by a fringe of fine, hair-like mastigonemes, by flap-like lips bounding the cytostomal opening, or by a short mobile rostrum (Brooker, 1971; Vickerman, 1976).

In *Rhynchomonas*, bacteria are taken in by phagocytosis through a permanent cytostome which is at the top of a proboscis, a mobile extension of the cytopharynx beside the anterior flagellum. The membrane-lined lumen of the proboscis forms a continuous tube with the cytopharynx down which bacteria and quite large particulate prey are conducted (Figure 9.6(b)). In an organism which lives as a coprophile, a scavenging proboscis attached to the anterior flagellum in such a way that it is moved by it, is an ideal implement for collecting food (Burzell, 1973). The microtubules in the walls of the cytopharynx have a cytoskeletal or a conducting function, according to Burzell (1975). This interpretation of functional morphology is not unlike those already described for the euglenoid flagellates, the suctorian *Dendrocometes*, or the ciliate *Nassula*.

Parasitic bodonids, such as *Ichthyobodo necator*, parasitize freshwater fish by attaching to the epithelial tissue of the gills of the host fish by a rostrum which penetrates and inserts an eversible pharynx. The scavenging function of the proboscis-rostrum is still maintained.

Trypanosomatids. The trypanosomatids, parasites with life cycles involving one (monoxenous) or two (heteroxenous) hosts, have evolved to occupy a wide variety of habitats (Table 9.1).

Those trypanosomatids which alternate between widely differing hosts, from a warm-blooded vertebrate to a cold-blooded insect, and from an aerobic environment richly supplied with soluble nutrients to an anaerobic insect gut in which the parasite has to work harder for its living, present an interesting morphological and biochemical study.

The structure of trypanosomatids is well-documented and need only be mentioned briefly. A single flagellum arises from a flagellar pocket whose position varies in different species and at different developmental

Table 9.1: Common Trypanosomatids and their Hosts.

Monoxenous genera:	Host(s):
Leptomonas	nematodes, insects, etc.
Crithidia	insects
Blastocrithidia	insects
Herpetomonas	insects
Heteroxenous genera:	
Trypanosoma	mammals/arthropods
Leishmania	mammals/arthropods
Endotrypanum	edentates/insects
Phytomonas	plants/insects

stages in the life cycle. In bloodstream trypanosomes the flagellum arises at the posterior end and extends forwards attached to the pellicle of the organism by a linear series of macular desmosomes (Vickerman, 1969a, b; Vickerman and Tetley, 1977). Overall body shape is maintained by a cytoskeleton of pellicle microtubules (Figure 9.7).

Polymorphism is common and can be related to cyclical changes in activity and metabolism as trypanosomatids pass through different stages of their life histories. Morphological changes are considered to be directly related to the re-positioning of the flagellum base-kinetoplast complex in relation to the nucleus (Vickerman, 1976). The most commonly recognized polymorphic forms are:

(1) *promastigote*, where the flagellum arises at the anterior end of the body;
(2) *amastigote*, in which there is no emergent flagellum;
(3) *epimastigote*, where the flagellum arises midway but anterior to the nucleus and is attached to the pellicle;
(4) *trypomastigote*, where the flagellum emerges from a flagellar pocket at the posterior end of the body and extends forwards, attached by an undulating membrane or by specific attachment points (Hoare and Wallace, 1966).

One can perhaps add to this list *staphylomastigotes*, amastigote-like forms of *Trypanosoma cruzi* which develop in tight clusters when cultured in the presence of triatomid bug embryo cells (Lanar, 1979). Perhaps it is more appropriate to designate staphylomastigotes as a style of growth rather than as a different polymorphic form. Figure 9.8 illustrates polymorphic range in trypanosomatids.

Figure 9.7: Body Organization in a Trypanosome. (a) A metacyclic trypomastigote of *Trypanosoma brucei* from the salivary gland of a tsetse fly. The single flagellum emerges from a posterior flagellar pocket and extends forwards adhering to the pellicle (x 22,000), (b) a transverse section shows the ring of microtubules (mt) internal to the pellicle which gives cytoskeletal support to the organism, and the close attachment of the flagellum (fa) (x 87,500).

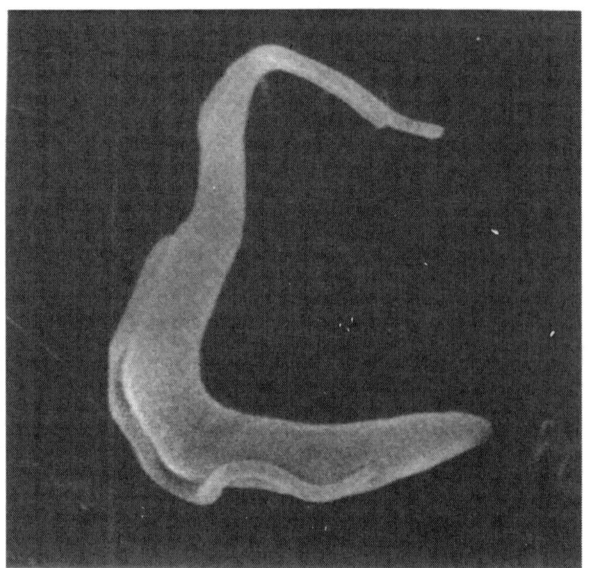

(a)

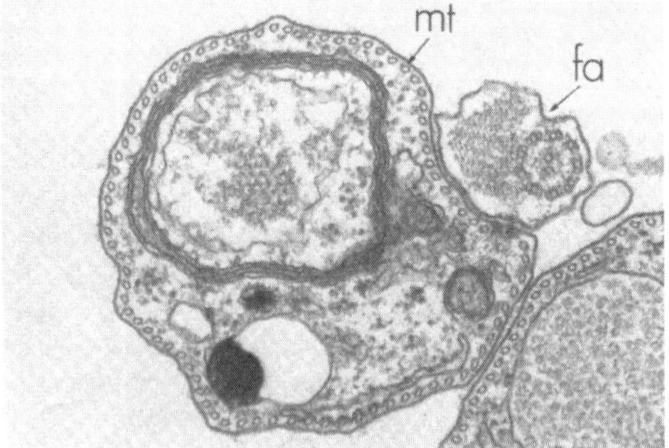

(b)

Source: Electron micrographs by K. Vickerman and L. Tetley, Glasgow University.

Figure 9.8: Polymorphism in Trypanosomatids. (a) Amastigote form; (b) promastigote form; (c) epimastigote form; (d) trypomastigote form.

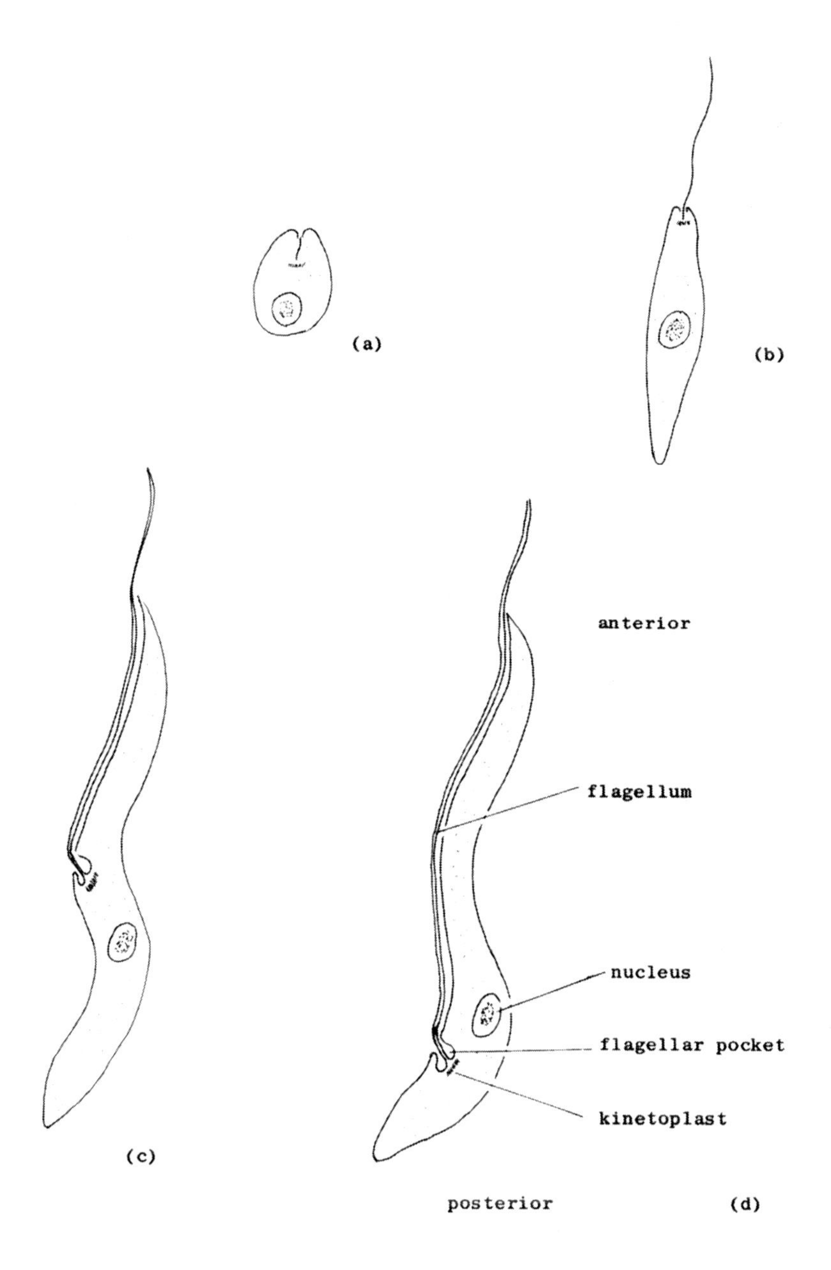

Genus Trypanosoma. Within this genus the multiplicity of minor variations in structure and biochemical reaction made it necessary to create subgenera of the mammalian species (Hoare, 1964). These subgenera fit into two major groups, Stercoria and Salivaria, which reflect differences in the life history and often in the biochemistry and physiology of the trypanosomes.

The Stercoria complete their life cycle in the hind gut of an insect and are transmitted by contaminated faeces to a mammalian host. Common stercorian trypanosomes are *T. lewisi*, non-pathogenic in rats and mice, *T. cruzi*, which causes Chagas' disease in humans, and *T. theileri* in cattle. Salivarian trypanosomes develop infective forms in the salivary glands of their insect host and are transmitted to the mammalian host by blood-sucking activity. For details of a typical life history consult Farmer (1980) or any other recent protozoology text. Trypanosomes of the *brucei* group, *T. b. brucei*, *T. b. rhodesiense*, and *T. congolense*, *T. evansi* and *T. vivax* are included here.

The site of food intake differs in the two groups of trypanosomes. Some stercorian trypanosomes do possess a cytostome-cytopharynx near the base of the flagellum in the epimastigote form although it has never been confirmed to function, for they lose it in the trypomastigote (*T. cruzi*). Others, e.g. *T. conorhini*, have a functional cytostome in the culture form only (Milder and Deane, 1969). A permanent cytostome opening near the flagellar pocket is present in culture forms of the non-mammalian trypanosome *Trypanosoma raiae* and is used for the intake of particles (Preston, 1969). Salivarian trypanosomes do not develop a cytostome-cytopharynx at any stage in their life history; they rely on pinocytotic uptake of macromolecules (Vickerman and Preston, 1976). Surprisingly this is not an all-over phenomenon. It is now well-established that uptake is restricted to the plasma membrane lining the flagellar pocket, although the initial binding of molecules may occur all over the surface of the flagellates. Acid phosphatases are also secreted into the flagellar pocket in some trypanosomes, which suggests that there may be extracellular digestion of small particles. A theory that the flagellar pocket represents a vestigial digestive sac has been proposed. Bristle coated vesicles lying beneath the walls of the flagellar pocket are implicated in the pinocytotic uptake of protein molecules, and may be the source of acid phosphatases. Unwanted material is also discharged from the flagellar pocket.

The metabolism of Trypanosomes. This has been the subject of several comprehensive works, and it seems superfluous to give anything other than the briefest outline here. For fuller information, reference should be made to von Brand (1973), Bowman and Flynn (1976) and Gutteridge and Rogerson (1979). Research on these organisms is also rapidly expanding into the field of immunology, where the presence or absence of variable antigenic surface coats at different stages in the life history is being recognized as an important factor for the survival of the parasite.

Trypanosomatids have a basically aerobic metabolism, although it may be considerably modified between the blood phase and the insect gut phase in heteroxenous species. A high rate of utilization of exogenous substrates is found, the uniformly acceptable energy-producing substrates being glucose, fructose and mannose. Table 9.2 makes a comparison of the carbohydrates used for energy production in different trypanosomatid genera.

In *Trypanosoma* species, differences exist in the metabolic pathways between bloodstream forms and culture and insect phases. Salivarian trypanosomes such as *T. b. brucei*, in an environment rich in exogenous glucose and dissolved oxygen as in the blood phase, has a rather abbreviated (and wasteful) oxidative metabolism involving incomplete oxidation of glucose to pyruvic acid and small quantities of glycerol. It lacks a cytochrome system and tricarboxylic acid cycle (Flynn and Bowman, 1973). Enzymes for this oxidative metabolic pathway in bloodstream trypanosomes are located in lysosomes or microbodies which act as compartments to enclose a multi-enzyme complex. Taylor and Gutteridge (1980) have shown that *T. b. brucei* glycosomes contain enzymes which initiate the glycolytic pathway, converting glucose to 3-phosphoglycerate and thus synthesizing high-energy ATP. Consequently it is a safe assumption that all the initial glycolytic enzymes are located in the glycosomes. *T. b. brucei* glycosomes do in fact contain hexokinase, phosphofructokinase, 3-phosphoglycerate kinase, aldolase and 6-phosphoglucose isomerase (Oduro *et al.*, 1980). *T. b. rhodesiense* shows a similar pattern in the trypomastigote (bloodstream) form, converting glucose to pyruvate, whereas the culture and insect phases use TCA cycle intermediates and complete oxidation to carbon dioxide (Ryley, 1962).

Stercorian trypanosomes, such as the non-pathogenic *T. lewisi*, and those of invertebrates have a cytochrome system and an operative TCA cycle and require proportionately less oxygen for glucose catabolism. Another stercorian trypanosome, *T. cruzi*, both as a blood trypomastigote and as an intracellular amastigote of reticulo-endothelial cells and muscle cells, follows this same metabolic pattern. *T. cruzi* has a functional TCA cycle and catabolizes the majority of the avail-

Table 9.2: A Comparison of Carbohydrates Used for Energy Production in Trypanosomatids.

	Glucose	Fructose	Mannose	Galactose	Sucrose	Raffinose
Crithidia fasciculata	+	+	+	+	+	+
C. oncopelti	+	+	+	–	–	–
Leptomonas collosoma	+	+	+	+	+	+
Leptomonas lactosovorans	+	+	+	+	+	+
Leishmania tropica	+	+	+	–	+	+
Lei. brasiliensis	+	+	+	–	+	+
Lei. donovani	+	+	+	–	+	+
Trypanosoma lewisi	+	+	+	–	–	–
T. cruzi	+	+	+	–	–	–
T. congolense	+	+	+	–	–	–
T. b. brucei	+	+	+	–	–	–

Note: The data have been collected from different workers from experiments carried out under different conditions but they do indicate distinct groups of metabolic behaviour. +, carbohydrate utilized for energy production; –, not utilized.

Source: Partly from Newton (1976).

able glucose to carbon dioxide.

Bloodstream trypomastigotes must prepare for change when they transfer to the midgut of their insect host. Transformation proceeds first to short stumpy forms in the blood, followed by activation of the mitochondrial respiration system. This allows fuller breakdown of glucose to carbon dioxide, acetate and succinate through cytochrome and the TCA cycle. Pre-adaptation to the more stringent conditions of the insect gut are essential (Vickerman, 1965).

Although the details of the oxidative metabolism of the bloodstream trypomastigote forms are well-established, information on epimastigote and metacyclic (infective) phases in the insect is only now emerging. The epimastigote must transform to an infective metacyclic form (a miniature trypomastigote) in either the insect salivary gland or hindgut, depending on whether it is a salivarian or stercorian trypanosome. The stercorian *T. cruzi* will transform to the metacyclic phase in response to extracts from recently-fed *Triatoma* (bed bug) intestine or stomach, but is unaffected by extracts of other tissues (de Isola *et al.*, 1981).

The transforming epimastigotes of *T. b. brucei* attach themselves to the lining of the insect salivary gland to avoid being expelled with saliva before they have acquired their antigenic surface coat (Vickerman, 1982). Anchorage is through hemidesmosomes (tentacular outgrowths) on the epimastigote flagellum and microvilli lining the salivary gland.

Genus Leishmania. This is a heteroxenous genus in which the protozoon is found as an amastigote in the tissues of man and other animals and as a promastigote in the midgut of an insect, usually a sandfly. Recently-developed methods for culturing *L. donovani* in mouse tumour cells will increase the scope of biochemical investigations. Amastigotes are engulfed by macrophage-like tumour cells in which they grow rapidly (Berens and Marr, 1979).

Leishmania species have adapted to metabolize a wider range of substrates than trypanosomes (Table 9.2) but they are not uniform in their metabolic pathways. *L. donovani*, in the intracellular mammalian stage, resembles salivarian trypanosomes in its lack of a functional TCA cycle and cytochrome system. The whole question of carbohydrate metabolism in *Leishmania* species has been reviewed by Marr (1980).

Leishmania tropica mexicana differs in that both the amastigote and promastigote stages have a classical terminal respiration cycle which is operated by enzymes concentrated into glycosomes. Carbohydrates (through glycolysis), amino acids and probably also fats (through β-oxidation) all pass into the TCA cycle before the terminal respiration cycle (Hart and Coombs, 1981):

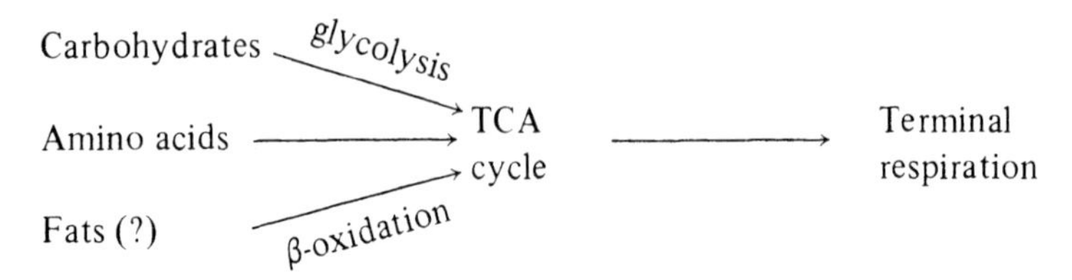

Other trypanosomatids. Crithidia and *Leptomonas*, both monoxenous in insects, have adapted to use a wider range of substrates than is found in the haemozoic trypanosomes, presumably in response to a less favourable environment (Table 9.2). The nutrition of *Crithidia fasciculata* is discussed in Hutner *et al*. (1979).

Leptomonas lactosovorans, a new species parasitic in reduviid bugs, described by Manaia *et al.* (1981), adds to the six substrates shown, lactose on which it can grow as effectively as on glucose, fructose and sucrose. The utilization of lactose as a growth substrate is unique amongst trypanosomatids. In common with the other lower trypanosomatids this *Leptomonas* can grow in a carbohydrate-free medium when it is thought to be utilizing glycerophosphate, amino acids and TCA cycle intermediates as alternative substrates. The rate of growth, however, is only one-third that achieved in an ideal substrate.

In this brief review of some trypanosomatids, carbohydrate metabolism in particular has been considered. It is worth re-emphasizing that this activity is directed entirely towards energy production (catabolic oxidation) and that no reserve carbohydrates are synthesized (or needed).

Diplomonadida

This order includes a variety of free-living and parasitic forms in which organelles tend to be paired and mirror each other. *Trepomonas agilis* traps bacteria in two cytostomal grooves lined by thin plasmalemma and with three oral flagella in each groove (Eyden and Vickerman, 1975). Food vacuoles containing bacteria are pinched off from the plasmalemma and circulate by active cyclosis, receiving enzymes in primary lysosomes produced on the cisternae of rough endoplasmic reticulum. This system of lysosome production is more reminiscent of ciliates than of flagellate protozoa.

The parasitic diplomonad *Giardia intestinalis* (*lamblia*) absorbs nutrients in pinocytotic vesicles through a ventral sucking disc. This sucking disc is a large structure supported by a coil of microtubules which attaches itself to host gut epithelial cells. Suction methods may vary between species, as some species leave behind a suction mark when detached whilst others leave no impression (Vickerman, 1982). Suction pressure is thought to be generated by the activity of two flagella which emerge from the posterior margin of the disc.

Since it is possible to maintain *Giardia* in axenic culture, some investigation has been made of its carbohydrate and energy metabolism. Although an anaerobic organism, it will also metabolize in the presence of oxygen, using a wide range of substrates. The initial steps are a glycolytic pathway similar to other protozoa, but the end-products of metabolism in both aerobic and anaerobic conditions are ethanol and acetate in addition to carbon dioxide (Lindmark, 1980). In other words the oxidation process is incomplete, which is not surprising in an organism with no mitochondria. There is neither a functional TCA cycle nor a cytochrome system.

Evidence from the carbohydrate metabolism of *Giardia* suggests a close biochemical relationship with the trichomonads.

Trichomonadida

The trichomonad flagellates are important parasites, although they are often more irritant than pathogenic. They inhabit the digestive tracts of invertebrates, and the intestine and urinogenital and oral cavities of vertebrates, feeding heterotrophically. Cannibalism occurs in some trichomonads (Mattern *et al*., 1969).

There is no functional cytostome nor even a cytostomal groove. A light area at the anterior end of the organism may represent a vestigial cytostome, but it has never been seen to function for the ingestion of food. Posterior ingestion of particulate material by invagination and pseudopodial outgrowths at no specific point or by pinocytosis is the general rule.

Food vacuoles, usually containing bacteria or other particles, are produced, the contents being mixed with acid phosphatases for the initial digestion process. Two types of lysosome-like vesicles are produced by the Golgi body: small granules (primary lysosomes?) containing acid phosphatase and larger lysosome vesicles containing neutral hydrolases. Unlike the trypanosomatids, carbohydrate reserves (large oval masses of glycogen) play a major role in the metabolism of trichomonads, forming 10-30 per cent of the dry weight of the organism.

A comparison of carbohydrate utilization in five species of tricho-

Table 9.3: Carbohydrate Utilization by Trichomonads.

	Pentoses	Glucose	Fructose	Galactose	Mannose	Maltose	Sucrose	Soluble Starch Glycogen
Trichomonas vaginalis	–	+	+	+	–	+	–	+
Tritrichomonas foetus	–	+	+	+	+	+	+	+
Trichomonas gallinae	–	+	+	+	–	+	+	+
Hypotrichomonas acosta	–	+	+	+	+	+	+	+
Tritrichomonas augusta	–	+	+	+	+	+	+	+

Note: A wide range of carbohydrates is utilized, with the exception of pentoses. The species listed inhabit a variety of vertebrate hosts (warm-blooded or cold-blooded) in which they are specific. +, carbohydrate utilized for energy production; –, not utilized.

Source: Partly from Shorb (1964).

monads shows obvious similarities to the trypanosomatids in that glucose, maltose, fructose and galactose are universally accepted. Mannose and sucrose and polysaccharide in soluble form, are also generally acceptable (Table 9.3).

In the zoomastigophoran flagellates, as in other groups of protozoa, information is accumulating on the carbohydrate metabolism of the organisms whilst lipid and protein metabolism remain much less well-known. It is established that *Trichomonas vaginalis* requires cholesterol as a source of sterols and the fatty acids, oleic, palmitic and stearic acid, when it is cultured in a defined medium.

Hypermastigida

Many hypermastigid flagellates have evolved an elaborate flagellar apparatus apparently employed to increase motility rather than to enhance food-trapping ability. This characteristic is particularly evident in those hypermastigid flagellates found in the gut of xylophagous insects such as termites, woodroaches and cockroaches. These symbiotic protozoa have a great deal of intrinsic interest, not the least of which is the acquisition, by some species, of ectosymbiotic bacteria to provide motility. This replaces conventional flagellar movement. Rod-like bacteria embed themselves end to end in specialized pockets on the host membrane in such a way that the bacterial flagella propel the host along in a gliding motion (Tamm, 1982). The host flagellum is unessential and can be removed without loss of motility.

Xylophagous insects are dependent on the cellulose-digesting activities of their symbiotic intestinal flagellate protozoa, which may account for one-seventh to one-third of the insect's total weight. It can be shown experimentally that the insects cannot survive without these flagellates unless their diet of wood fragments is replaced by predigested cellulose. A typical population of flagellates would normally include species from the orders Oxymonadida, Trichomonadida and Hypermastigida, but only certain hypermastigid flagellates, *Trichonympha* in particular, have been shown to be essential. In 1934, Cleveland and co-workers described the flagellate fauna and the nature of the relationship between the protozoa and the woodroach (*Cryptocercus*) in a detailed monograph. This is still the most comprehensive definitive description available.

Trichonympha campanula has no operative cytostome, and yet a large part of the body is filled with wood fragments. This is another example of phagocytosis by pseudopodial action. Behind the elaborate flagellated anterior region of the body, the posterior end remains relatively unorganized, soft and deformable, with a sticky surface membrane to which wood fragments adhere. Cleveland (1925) has described the way in which *Trichonympha* ingests its food. The body is seen to shorten due to contraction of oblique and longitudinal fibrils, which creates a temporary posterior concavity into which wood fragments (and other particles) are drawn. Accidental capture of small protozoa does occur. As the fibrils relax, the undifferentiated cytoplasm flows backward, surrounding the adhering material which is absorbed into the organism. *Trichonympha* and *Trichomitopsis* (a trichomonad flagellate of termites) depend exclusively on cellulose as a carbon source, and they are believed to secrete a cellulase for its breakdown. Although it has not been established beyond all doubt that cellulase is produced by the protozoa, the contribution of cellulolytic bacteria is still uncertain. The flagellates are obligate anaerobes with a carbohydrate metabolism probably not unlike that of ciliate protozoa of ruminants. Anaerobic fermentation follows the scheme: cellulose (or hemicellulose) → cellobiose → glucose → acetic acid, carbon dioxide and hydrogen. These latter are the chief metabolic end-products. Production of acetic acid and carbon dioxide accounts for 75 per cent of the carbon in the cellulose digested. The flagellates of the wood-roach produce traces of pyruvate, succinate and lactate in addition to acetate. Continuing the analogy with ciliates in ruminants, acetic acid released into the host hindgut forms the principal energy source of the insect (Hungate, 1943b). In common with other zoomastigophora, polysaccharide reserves are glycogen, stored in the form of small granules. The flagellates are not thought to provide a nitrogen source for the insects. The whole question of the biochemistry of the relationship between these flagellates and their insect hosts has been admirably reviewed by Honigberg (1967).

Opalinata

The opalines are common inhabitants of the rectums of frogs and toads where they live as endocommensals. Not so common, but certainly established, is their occurrence in the posterior alimentary tract of fish. This is a small subphylum of only four genera, *Opalina*, *Cepedea*, *Zelleriella* and *Protoopalina*, which have been fully described by Wessenberg (1978). As a group they are remarkably similar, all possessing a complete covering of short fine cilia and lacking a mouth. The principal characters used in identification are the number of nuclei and the cross-sectional shape, whether circular or much flattened. *Opalina ranarum* is the best-known member of the group. Its large flattened shape and numerous small nuclei make it outstandingly different from other protozoa.

Opalinids have found themselves a very convenient and comfortable niche to occupy. In the frog rectum they have abundant food and shelter beyond the reach of the host enzymes. They absorb unwanted nutrients and apparently have no effect on their host even when present in very large numbers.

Food is taken in by pinocytosis at sites between the pellicular folds. Large numbers of endocytotic vesicles accumulate in an orderly fashion lined up in a framework of microfilaments in the cortical zone of the cell. Noirot-Timothée (1966) described two types of vesicles arranged in alternating rows perpendicular to the pellicular folds. Spherical rough-coated vesicles alternating with smooth vesicles of variable shape were thought initially to be bringing different nutrients into the organism.

By use of radioactive tracer it has been shown that the rough vesicles represent a temporary stage in the feeding process (Munch, 1970). The rough coat soon disappears as these vesicles fuse into larger ectoplasmic alveoles, perhaps more appropriately described as digestive vacuoles. Golgi vesicles bring enzymes to them and as digestion proceeds the halo of small vesicles, which has come to be associated with late digestion vacuoles, appears round them. This may take one day or more. Egestion vacuoles can also be recognized in the cortex of the organism.

Although the Opalinata are currently classified as a subphylum of the Sarcomastigophora, Wessenberg (1978) considers that they are sufficiently different to justify a separate phylum. This view has not been adopted by Levine *et al.* (1980) in revising the protozoa, for in nuclear structure and division and in gamete production they are close to the Mastigophora. Distinctive features are the mode of asexual reproduction by cleavage into several small trophonts, still retaining the multinucleate condition, and the initiation of a sexual phase under the influence of the host's own sex hormones (Wessenberg, 1961).

In Summary

This chapter started by recognizing that the Mastigophora encompassed a wide variety of nutritional types. The distinction between the plant-like and the animal-like flagellates has been stressed and the more important orders of each group have been described. One factor which emerges is that there is a surprising uniformity in the metabolism of the parasitic zoomastigophora of vertebrates, although their morphological characteristics separate them into different orders. The absence of carbohydrate reserves is the result of living bathed in nutrient fluids, for free-living flagellates produce a range of polysaccharide stores.

Feeding in the Opalinata is by pinocytosis as these endocommensals of amphibia are without cytostomes. The whole range of digestive vesicles is located in the cortical zone of the organism.

10 SARCODINA

The Sarcodina are the amoeboid protozoa which trap food by some form of pseudopodial action, pseudopodia which range from the single blunt lobopodium to the delicate tracery of reticulopodia which project from a foraminiferan shell. Both are designed to perform the same function, to trap food. Sarcodines have no cytostomes and have bodies of a less organized shape than most protozoa. Shape comes as a result of being encased in an elaborate shell or test, which imparts species characteristics.

Most free-living sarcodines are carnivorous; only a few are herbivores or browsers. Even the carnivores do not actively seek prey, but respond to chance contact from a food organism by producing pseudopodia. Absorption of nutrients by pinocytosis is an equally important activity, particularly amongst parasitic amoebae. Much of the recent research on amoebae has been directed towards investigating the nature of the cell surface, how parasitic amoebae attach to host cell surfaces and how free-living amoebae respond to contact by prey organisms.

There are two major groups of the Sarcodina, the superclasses Rhizopoda and Actinopoda, which reflect the type and form of action of pseudopodia. The Rhizopoda include the classes Lobosea, Acarpomyxea, Acrasea, Eumycetozoea, Plasmodiophorea, Filosea, Granuloreticulosea and Xenophyophorea. The relative obscurity of some members of these classes determines the amount of information which is available. Most common amoebae belong to the Lobosea, producing lobose or at least broad pseudopodia, which may project from a test or shell in the testate amoebae (Figure 3.3). Amongst the other rhizopodans the Granuloreticulosea probably figure next as they are known for their elaborate chambered shells and more recently for the interest which has focussed on their symbiotic partnerships with algae (Lee, 1980b).

The Actinopoda are generally free-floating, planktonic organisms where a spherical shape and arrangement of radiating axopodia confer a definite advantage. Axopodia differ from other pseudopodia in being long and slender and possessing microtubular cytoskeletons which have a permanence not found in other types of pseudopodia. The classes included are Acantharea, Polycistinea, Phaeodarea and Heliozoea.

Endocytosis in Amoebae

The large naked amoebae which produce one to several lobose pseudopodia are ideal subjects for studying phagocytosis, especially as they tend to take quite large prey, perhaps a paramecium or a tetrahymena.

The simplest version is seen in those amoebae such as *Chaos carolinense*, a large multinucleate carnivore which has a preference for *Paramecium* as food. *Chaos* will produce a food cup in response to contact by live or freshly-killed *Paramecium* provided that contact is one second or longer. The sequence of events involves gentle contact when an inducer, a labile substance probably located on the cilia, binds on to specifically charged sites on the mucopolysaccharide slime coat or glycocalyx that is characteristic of large amoebae (Christiansen and Marshall, 1965). As the paramecium pulls away, clumps of mucopolysaccharide granules remain attached to its cilia at the point of contact, causing clumping of the cilia. This has the effect of upsetting the balance of the ciliate and making it gyrate back towards the amoeba, so the probability of capture is increased.

The production of food cups by upsurge (circumvallation) or by invagination involves pseudopodial outgrowths which vary with different species of amoebae. The reaction must be rapid enough to ensure successful capture; in *Chaos* a food cup may take from two to twelve seconds to complete. The point of contact on the plasmalemma forms the fixed centre of gelled protoplasm around which a ring of plasmalemma swells up, the lips converging over the prey and enclosing it in a vacuole. Ultrastructural studies have shown that the formation of the fixed centre of the food cup is due to a specialized region below the point of contact in which a fibrillar network and deep invaginated canals develop. Complete separation of the vacuole and sealing of the plasmalemma at the point of invagination may take as long as ten minutes, during which time the ciliate continues to swim. The food vacuole shrinks, losing water perhaps through the deep canals which are invaginated for this purpose, until it assumes the contours of the now dead paramecium (about 15 minutes).

Chaos will select *Paramecium* in preference to *Tetrahymena* or *Euglena gracilis* when presented with a mixed culture of these organisms. By increasing the numbers of *Tetrahymena* and *Euglena* in the culture, feeding on *Paramecium* is inhibited (Lindberg and Bovee, 1976). Perhaps too much surface contact by unwanted organisms, especially other ciliates, disturbs the amoeba's ability to respond. This is confirmed when batches of detached *Tetrahymena* cilia added to a *Chaos* culture inhibit the phagocytosis of *Paramecium* as effectively as live *Tetrahymena*. The active agent seems to be some substance on the surface of the cilia. This surface reaction can be used to induce *Chaos* to cannibalize its own kind. Dead *Chaos* coated with cilia from *Tetrahymena* will provoke a feeding response in *Chaos*; the dead amoebae are cannibalized although normally expelled 30 minutes later without being digested (Lindberg and Bovee, 1976).

Roth observed feeding in the carnivorous *Chaos* (*Pelomyxa carolinense*) and followed progressive breakdown of organelles of the prey organism *Tetrahymena*, starting with the pellicle membranes, mitochondria, nucleus and finally cilia and trichocysts (Roth, 1960). The age and state of digestion of food vacuole contents can be gauged from the morphological changes of the prey organism that has been captured.

Other amoebae have different food preferences and characteristics. The smaller *Amoeba proteus* may engulf 20 *Tetrahymena* in 24 hours. It also shows a definite preference for some ciliates with *Tetrahymena* in top position. Small soil amoebae such as *Acanthamoeba* and *Naegleria* feed by phagocytosis of bacteria. The herbivorous *Pelomyxa* ingests non-motile algae and diatoms at its posterior end (uroid) by gradual invagination, as the food is often much longer than the body of the amoeba.

Once food capture is completed, the food vacuole in an amoeba separates from the surface plasmalemma and passes through a series of changes similar to those described in Chapter 6, starting with death of the prey due to dehydration and followed by its gradual digestion over several hours. Egestion vacuoles may still contain undigested protein material although polysaccharides are digested completely.

The relationship between vacuoles and vesicles in amoebae is represented in a generalized diagram in Figure 10.1. As in other protozoan groups, acid phosphatase production is a direct feeding response. With the exception of newly-formed ones, food vacuoles of all ages contain large quantities of acid phosphatase, regardless of the nutritive value of the contents. Table 10.1 lists the major enzymes which have been confirmed in food vacuoles or lysosomes by histochemical staining methods or from fractionated cell homogenates in selected free-living amoebae.

The two endocytotic processes, phagocytosis and pinocytosis, are closely linked and must reach a balance in the amoeba. If the rate of phagocytosis increases then the rate of pinocytosis must decrease proportionately for the total endocytotic compartment is critical. In *Acanthamoeba* the total endocytotic volume may not exceed 15 per cent of the cell volume (Bowers, 1977).

Figure 10.1: Endocytotic Vacuoles and Vesicles in Amoebae. Nutrients enter by phagocytosis or pinocytosis, then follow an identical sequence of events: DVI, DVII (receiving enzymes from primary lysosomes), DVIII (pinching off secondary vacuoles), egestion vacuole. For a fuller explanation refer to Chapter 6.

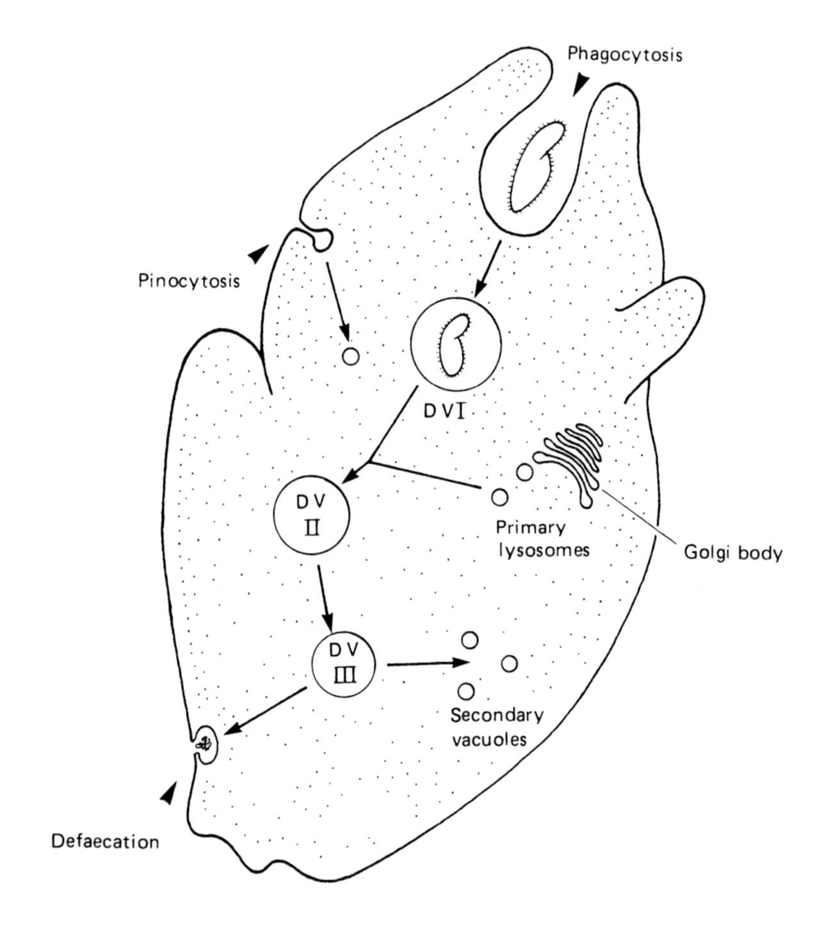

Table 10.1: The Major Digestive Enzymes in Free-living Amoebae.

Enzyme	Optimum pH	Species
Acid phosphatase	5.0-6.5	*Amoeba proteus*
	4.5-5.0	*Chaos carolinense*
	5.0	*Acanthamoeba* sp.
Acid protease	3.8-4.0	*Amoeba proteus*
	3.5-4.0	*C. carolinense*
	4.0	*Pelomyxa palustris*
	3.0	*Acanthamoeba* sp.
Acid phosphomonoesterase	4.5	*C. carolinense*
α-glucosidase	4.0	*Acanthamoeba palestinensis*
	3.8	*C. carolinense*
	4.0	*Acanthamoeba* sp.
β-glucosidase	3.6	*A. palestinensis*
	4.0	*Acanthamoeba* sp.
Amylase	6.0	*Amoeba proteus*

Note: Enzymes identified by histochemical means or from fractionated cell homogenates, in food vacuoles and in lysosomes and therefore assumed to be associated with digestion of ingested food material.

Pinocytosis is an active process which is as useful in providing nutrients for the metabolism of the organism as is phagocytosis. The characteristic rosette shape which is indicative of an amoeba in active pinocytosis was first noticed by Mast and Doyle (1934), but it was Chapman-Andresen (1962, 1967) who recognized that active pinocytosis results from the presence of certain inducer substances in the external medium. This has been discussed more fully in Chapter 5.

Rhizopoda

This superclass includes a wide variety of amoeboid organisms which move and feed by producing blunt, filiform or reticulate networks of pseudopodia or by a simple flowing action.

Lobosea

The many small amoebae of soil and water environments which have already been mentioned in Chapter 3, are lobose amoeba. Some have become facultative parasites, producing a range of effects from harmless, to the merely troublesome, to fatal infections.

Small Lobose Amoebae

The small size and relatively unselective feeding habits of many soil amoebae have made the move from free-living to facultative parasitism or endocommensalism easy to achieve. *Naegleria* and *Acanthamoeba* are two genera which have been positively identified with this habit, as pathogens and as harmless invaders. A review of the biology of these small amoebae may be found in Schuster (1979).

The possession of a flagellated stage in *Naegleria* separates it from the other common small amoebae, *Hartmanella* and *Acanthamoeba*, but identification must be taken further than this. The importance of *Naegleria* as a human pathogen has made recognition of individual species of crucial importance. Of the four species of *Naegleria* described in the literature only *N. fowleri* is a pathogen, invading the human body by way of the nasal passages during swimming in hot-water pools. Migrating to the brain and anterior spinal cord, it multiplies between the meninges and causes fatal amoebic meningoencephalitis. *N. fowleri* forms particularly resistant cysts which can survive very low and very high temperatures; thus as an infection in water it is very difficult to eradicate.

N. gruberi is a common non-pathogenic soil and water amoeba which produces cysts as readily as *N. fowleri* and both transform to flagellated organisms in water. Their feeding preferences differ, but not consistently enough to be used as a sound diagnostic character. *N. gruberi* favours phagocytosis and is a more discriminating feeder than *N. fowleri*, which can obtain most of its nutrients by pinocytosis.

The harmless *N. gruberi* can be distinguished from the pathogenic *N. fowleri* by the use of Concanavalin A-induced agglutination tests. A concentration of 1×10^6 amoebae produces a maximum response to Con A (up to 100 μg/ml), when *N. gruberi* will clump and *N. fowleri* will not (Josephson *et al*., 1977). Recently isoenzyme electrophoresis has been used to distinguish between morphologically similar strains of these two species. Thirty separate strains were identified of which 22 were high-temperature pathogenic strains of *N. fowleri*; three were high-temperature non-pathogens and the remaining five were non-pathogenic strains of *N. gruberi* (Nerad and Daggett, 1979).

Species of *Acanthamoeba* are very common soil and water amoebae which ingest nutrients by phagocytosis and pinocytosis. Acanthamoebae are very voracious bacteria-feeders although they may exercise some selection. The importance of binding bacteria to the surface of *Acanth-*

amoeba as a prelude to moving them to the uroid for engulfment has been discussed in Chapter 5. Although no specialized organelles for food uptake exist, as the amoeba moves, its sticky posterior end draws the bacteria into a single large food cup (Page, 1967). Fine acanthopodia (filamentous projections) which appear on the advancing pseudopod may pass backwards and help in food uptake.

A. castellani is readily grown in a chemically-defined medium which contains 21 amino acids, with glucose and acetate, when a maximum growth rate of 14 to 20 hours generation time is achieved. Reducing the number of amino acids to nine essential ingredients, near maximum growth still results. Growth may still continue when the medium contains only six selected amino acids (L-arginine HCl, glycine, L-isoleucine, L-leucine, L-methionine and L-valine) with glucose and citric acid, although the generation time will fall to more than 60 hours (Byers *et al.*, 1980).

Cyst formation in *Acanthamoeba* is an important factor in allowing the organism to establish in a wide variety of environments, including the warm-water discharges of factories. It is also becoming a human pathogen with habits similar to *Naegleria*, so an understanding of the conditions which induce cyst formation is necessary. Synchronous encystment may be artificially induced in culture by depriving *A. castellani* of specific carbon sources, glucose and acetate, although not by general starvation. *A. palestinensis* will encyst in various iso-osmotic solutions including NaCl, KCl, $MgCl_2$, $CaCl_2$, glycine or sucrose.

Other Naked Amoebae

Certainly one of the best known freshwater amoebae is *Amoeba proteus*, one of four recognized species of *Amoeba*. It is the subject of an extensive monograph by Jeon (1973) and no detailed description is necessary here. Of interest is the fact that certain strains of *A. proteus* contain symbiotic bacteria. *A. proteus* strain D has an obligate population of symbiotic bacteria which disappear from the amoeba at temperatures above 26.5°C, with consequent death of the host. The precise metabolic relationship between these amoebae and their symbionts has not been elucidated although ultrastructural study gives some indication. At temperatures above the critical point, the symbiotic bacteria are actually digested, thus removing any continuing contribution which the symbionts may have been making towards the metabolism of the amoebae.

The Endamoebidae

Many small amoebae are obligate parasites and those of the family

Endamoebidae are well researched, particularly because of their connection with man. This family contains parasites and commensals of the digestive tract and other cavities of many vertebrates and arthropods. *Entamoeba histolytica* is parasitic in the digestive tract of man, often causing severe and debilitating amoebic dysentery. *E. coli*, commensal in the digestive tract, and *E. gingivalis*, commensal on the gums and tonsils, have long been recognized as distinct species.

Phagocytosis and pinocytosis are not markedly different from similar activities in free-living amoebae, although the choice of food differs. Fragments of host cell tissue or nuclei, erythrocytes, starch grains or bacteria are engulfed into food vacuoles for digestion. *E. coli* restricts its feeding to the contents of the host intestine without damaging the lining epithelium; thus it remains harmless. *E. histolytica* releases lysosomes containing proteolytic enzymes at its surface membrane which help to lyse the tissue lining the gut (Martinez-Palomo, 1982). In order to do this efficiently *E. histolytica* must be able to attach to the surface of the gut epithelial cells. By experiment in culture it has been found that *E. histolytica* requires cysteine and ascorbic acid before it can attach fully extended on surfaces; without these substrates it rounds up and often lyses.

The trophozoite of a healthy *E. histolytica* has a surface which is convoluted or at least wrinkled, and produces a number of very long, fine filopodia. Lushbaugh and Pittman (1979) believe that these filopodia have a variety of functions – in endocytosis and pinocytotis, in exocytosis, in attachment to surfaces, in the penetration of tissues, in the release of cytotoxic substances and in contact cytolysis of host cells.

Entamoebae of the gut, being in anaerobic conditions, show metabolic pathways similar to those of rumen ciliates previously described, although it should be recognized that carbohydrates of plant origin would not normally be available in quantity. The capacity to exploit the parasitic situation by rapidly converting exogenous carbohydrate to reserve glycogen is a parallel to that of rumen ciliates.

E. histolytica has a limited tolerance for oxygen in its environment and will even utilize it, provided it does not exceed 5 per cent, and provided reducing agents such as cysteine are present. The importance of cysteine in this instance is to detoxify the products of oxygen reduction which could otherwise accumulate. Thus cysteine, also a protein constituent, has more than one function for gut amoebae. For axenic growth, carbon dioxide is necessary, but excess hydrogen is inhibiting (Band and Cirrito, 1979). The addition of yeast extract improves growth in culture, but it does not spare the carbon dioxide requirement. As carbon dioxide is normally present in the intestine it is not surprising that *Entamoeba* has come to use it.

Carbohydrates utilizable for growth of *E. histolytica* are few. Glucose is the principal sugar required and at the level of the large intestine where this entamoeba normally lives, glucose would not be freely available. This is considered to be a possible reason why *E. histolytica* invades the epithelial lining of the gut. To ensure that sufficient glucose is available for growth *E. histolytica* operates a specific membrane transport system which is one hundred times more efficient than uptake of glucose by endocytosis (Serrano and Reeves, 1975). The metabolic pathways for the utilization of glucose are rather different from those in other protozoa, as *E. histolytica* possesses neither mitochondria, nor Golgi body, nor does it operate a TCA cycle. The end-products of glucose catabolism are the same whether in aerobic or anaerobic conditions: ethanol, acetate, carbon dioxide and hydrogen. Aerobic breakdown produces a higher proportion of acetate to ethanol (3:1), whilst anaerobic breakdown is the reverse, acetate to ethanol (1:3).

One point which emerges from biochemical studies on *Entamoeba* is that some of its metabolic pathways have alternative glycolytic enzymes and slightly altered routes. This lack of specialization in enzyme systems suggests that the entamoebae are not as evolutionarily advanced as most protozoa and that they resemble bacteria in some ways (Weinbach *et al.*, 1976; Wood, 1977).

Granuloreticulosea

The amoebae of this class are characterized by the possession of fine hyaline pseudopodia forming a network of non-anastomosing strands. Apart from a few lesser known genera the remainder are the foraminiferans which secrete tests of one or many chambers. Unfortunately empty tests are a more familiar sight than live organisms (Figure 10.2).

In life the foraminiferan uses its extended pseudopodial network (reticulopodia) for trapping food. The benthic existence which most species lead means that elaborate devices can be developed for collecting bacteria, algae, particularly pennate diatoms, and organic detritus. *Astrorhiza limicola* creates a spider's web of fine pseudopodia; *Bathysiphon* has pseudopodia which terminate in sticky knobs; *Pilulina* shapes its shell into a pit into which food organisms fall; and *Elphidium crispum* weaves itself into a feeding cyst formed from the pseudopodial network. Prey is trapped and engulfed by a flowing-around process involving a section of the pseudopodial network.

Figure 10.2: Foraminiferan Tests. (a) *Discorbina globularis;* (b) *Elphidium crispum*, (c) *Uvigerina angulosa*; (d) *Spiroloculina limbata*, (e) *Peneroplis pertusus*; (f) *Rotalia baccari*.

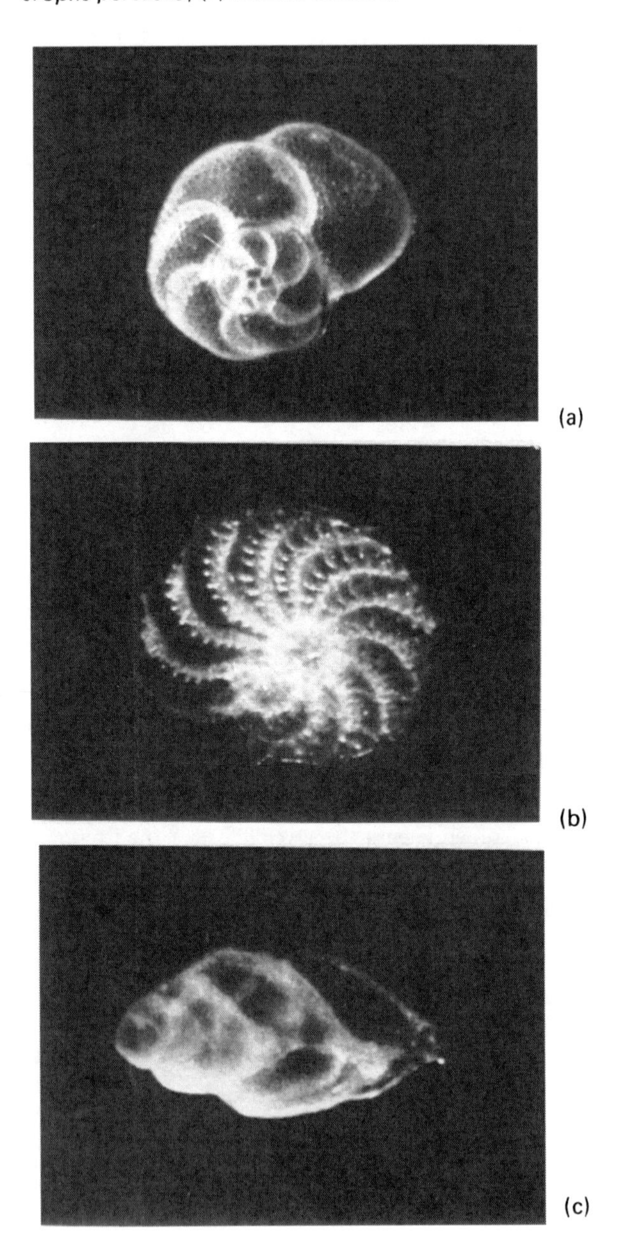

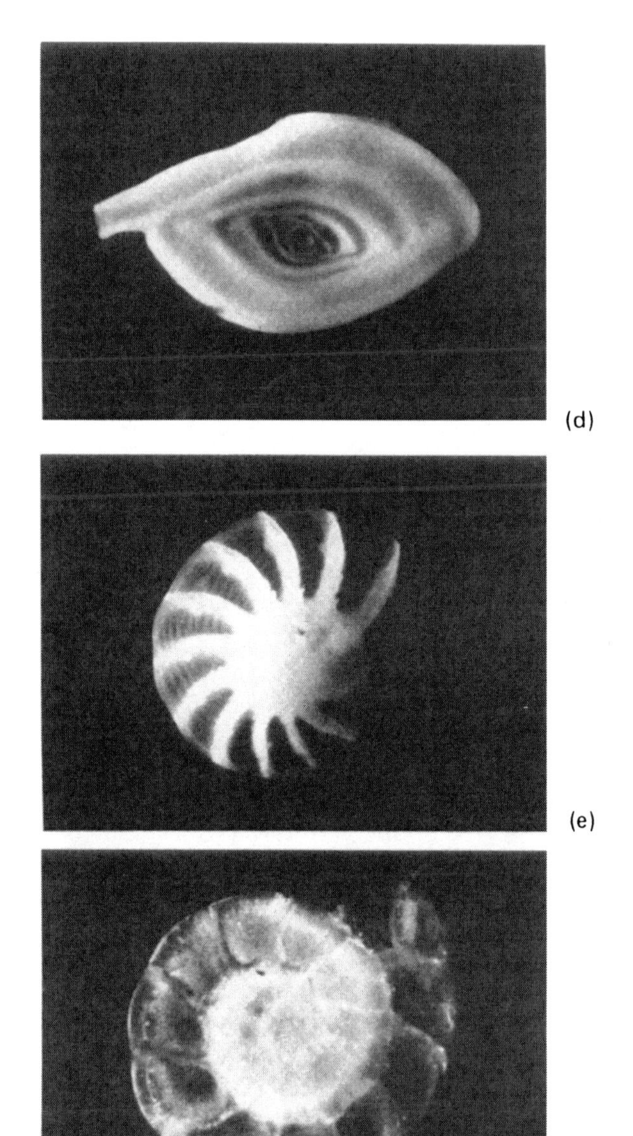
(d)
(e)
(f)

Deposition of a protective shell is an important activity for the growing foraminiferan. It requires energy. Growing juvenile *Allogromia* eat 200 per cent more food organisms than mature forms (Lee *et al.*, 1966a). Tropical and subtropical foraminiferans grow very large and often in areas where the environment is low in nutrients. This is only possible because they establish a symbiotic relationship with certain photoautotrophic algae much the same as that between corals and their zooxanthellae. Large tropical foraminiferans contain symbiotic diatoms, green algae and zooxanthellae which are believed to help in forming the shell. The many unanswered questions, including that of the interaction between the carbon budgets of the two symbiotic partners, and the interesting idea that diatoms are 'farmed' by their foraminiferan partners, are reviewed by Lee (1974, 1980b) and Lee *et al.* (1965, 1966a).

Actinopoda

The Actinopoda include the most beautiful of the sarcodines, radiolarians (Polycistinea and Phaeodarea) with skeletons of silica set into elaborate shapes, and heliozoans with axopodia radiating from all sides and earning them the name of sun animalcules. The Acantharea have more geometric patterns of spines of strontium sulphate (Figure 2.8(c)). Almost all have adapted for the planktonic way of life which is typical of the superclass.

Radiolaria

Radiolaria, by common usage, is a convenient collective term for the many marine forms which float around in the plankton and trap food in rhizopodial networks. The body of a radiolarian consists of a central capsule where lipid and carbohydrate are stored after digestion. Enclosing this is a capsule membrane and a skeleton of spicules often formed into an elaborate shape with pores through which the extracapsular cytoplasm extends before spreading out as a rhizopodial net (Figure 10.3).

Radiolarians tend to be omnivorous, trapping any suitably small organisms found in the plankton, such as diatoms, colourless flagellates, ciliates and small crustacea. The prey is captured by cytoplasmic streaming involving axial microtubules in the rhizopodia although the capture strategy varies with different prey (Anderson, 1980). Algae are quickly immobilized and adhere to the rhizopodium before being

Figure 10.3: Radiolarian Skeletons Form into Elaborate Shells, Lattices or Networks of Silica Spicules.

engulfed into a food vacuole. Food vacuoles are moved to the sarcomatrix surrounding the central capsule for digestion. If a crustacean is trapped, its struggles stimulate active cytoplasmic streaming in the rhizopodial network. Rhizopodia wrap round the appendages of the prey, exerting increasing pressure until the crustacean exoskeleton is ruptured and pieces of tissue are released (Figure 10.4).

In common with other marine sarcodines radiolarians have their population of algal symbionts. These zooxanthellae are enclosed loosely in the rhizopodial network or in a cytoplasmic envelope, but there appears to be no direct mutual exchange of nutrients. As superficial symbionts they photosynthesize and divide to keep pace with loss, for when they come to lie near the sarcomatrix they are digested. The intriguing idea that radiolaria culture their own symbionts and can regulate the numbers which occupy the rhizopodium is put forward by Anderson (1980). The colonial *Sphaerozoum* controls its numbers at 30-50 symbionts per cell.

Heliozoea

Heliozoans usually inhabit fresh waters except for the minute *Oxnerella* which creeps around on estuarine mud flats, ingesting bacteria and

Figure 10.4: A Radiolarian Ingesting Active Prey. (a) A crustacean larva is caught in the rhizopodial network; (b) a rhizopodium encloses it; (c) its struggles stimulate a crushing rhizopodial flow, the exoskeleton is ruptured; (d) part of the prey contents are separated and enclosed in a food vacuole which moves in towards the central capsule (CC)

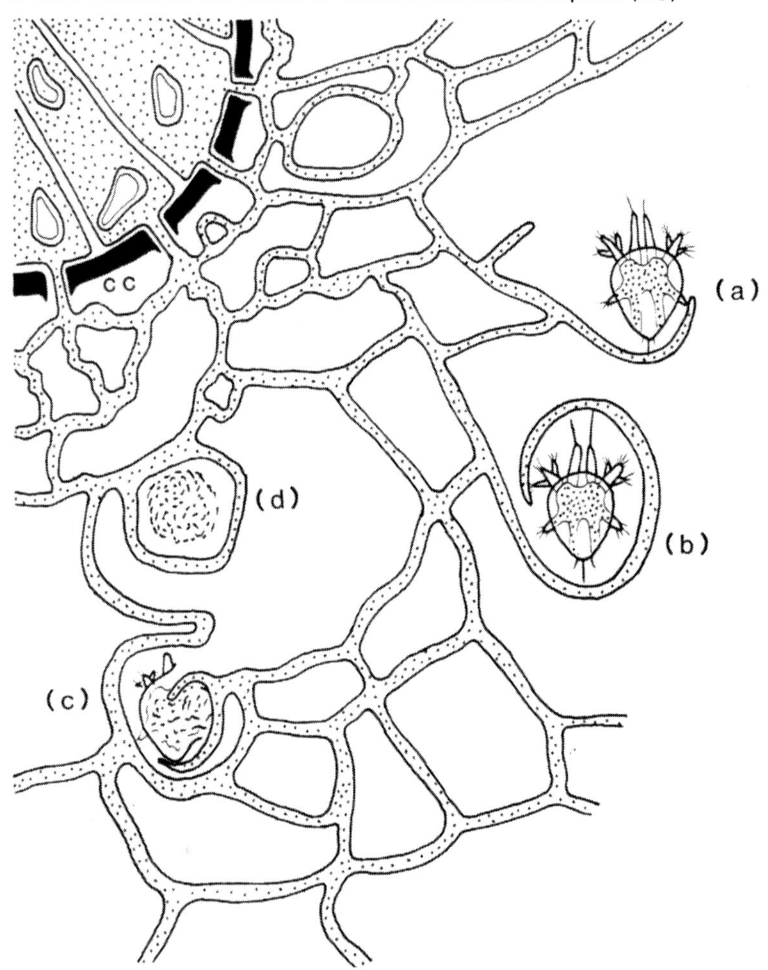

Source: Based on Anderson (1978), by permission of Longman Group Ltd.

small unicellular algae through its long, fine axopodia. Although *Oxnerella* has not been reported from many localities, it is probably more common than is realized. Its inconspicuousness makes it easy to overlook in a sample of mud. The larger heliozoans, in spite of a transparent appearance and lack of motility, are conspicuous organisms.

Phagocytosis in the large *Actinosphaerium* involves several axopodia in ingesting active food organisms. The prey makes contact accidentally, which stimulates the tips of the axopodia to produce a paralyzing secretion. A non-ingesting axopodium contains an axial rod of a double coil of microtubules which give it a semi-rigid structure. To allow the axopodia to function as ingestion pseudopodia the axial rod of microtubules breaks down temporarily and is reabsorbed into the cytoplasm. Adjacent axopodia may then co-operate by flowing round the prey, enclosing it in a phagocytotic vacuole which is withdrawn into the body of the heliozoan.

Figure 10.5: *Actinophrys*. The trophic organism shows axopodia with a core of microtubules (MT) and superficial extrusomes (EX) which move up the axopodium during feeding to fuse with the ingestion pseudopod. FV, food vacuole; VS, vesicles involved in digestion; CV, contractile vacuole.

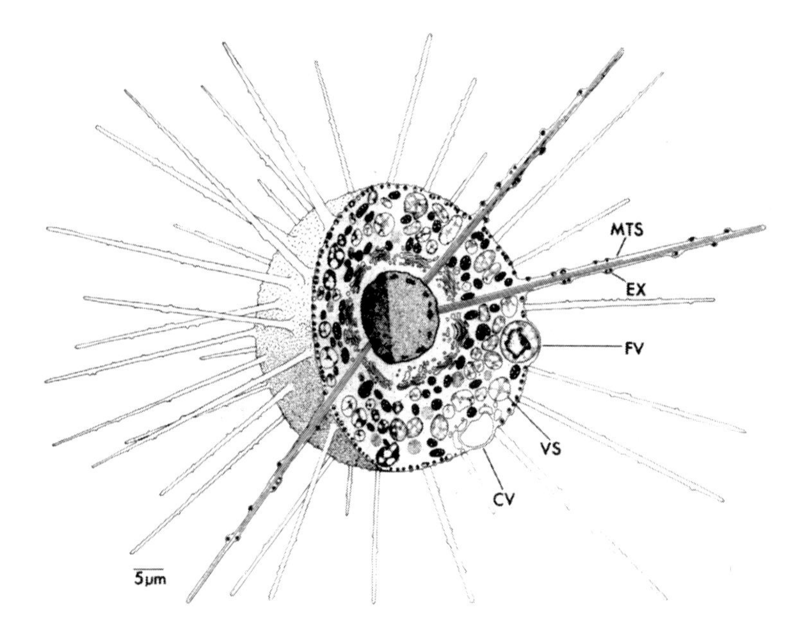

Source: Drawing by D.J. Patterson, University of Bristol, by permission.

Figure 10.6: Ingestion by Funnel Pseudopod in *Actinophrys*. Inset shows ciliate *Colpidium* caught on an axopodium; pseudopodial arms grow out and expand into a funnel pseudopod which encloses the prey organism. A light micrograph of *Actinophrys* with *Colpidium* in the final stage of engulfment. Details of the prey mouth and cilia are still visible (x 1100).

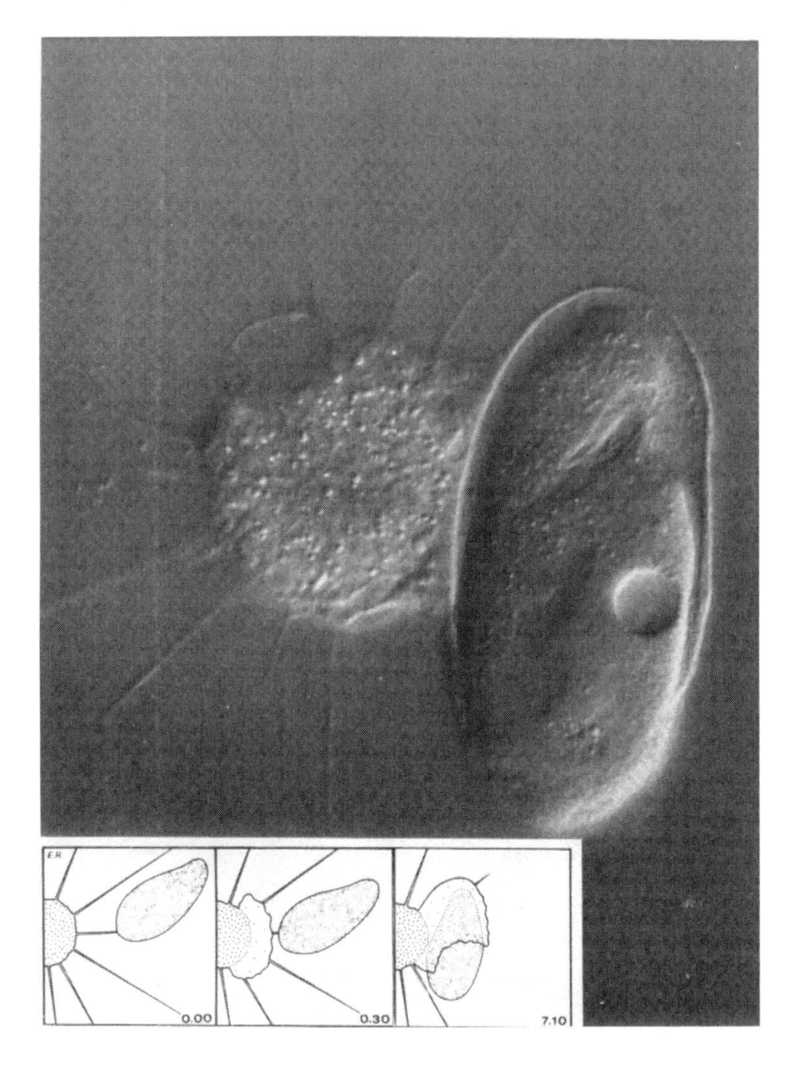

Source: Drawings by E. Hausmann, from Hausmann and Patterson (1982); photograph by D.J. Patterson, University of Bristol, by permission.

Actinophrys sol is a small, free-floating freshwater heliozoan which can build up large populations if the appropriate food organisms are present (Figure 10.5, and Patterson, 1979). It captures and engulfs prey as large as itself. When a prey organism, for example *Colpidium*, makes contact with axopodial arms it is drawn in towards the predator by arm resorption. Within five minutes of capture fine pseudopodial arms start to grow out, coalescing to make a deformable funnel (Patterson and Hausmann, 1981; Hausmann and Patterson, 1982). This pseudopodial funnel engulfs the prey, which after 20 minutes is enclosed in an endocytotic vacuole for digestion (Figure 10.6). Small bodies, extrusomes, lie below the plasmalemma and travel along the axopodial arms during feeding. Their involvement in feeding is indisputable, for their numbers (up to 10,000) are reduced considerably during feeding. Two functions for extrusomes are possible: the production of adhesive material and replenishment of membrane used in phagocytosis, as extrusomes fuse with the plasma membrane of the funnel pseudopod (Hausmann and Patterson, 1982). The whole feeding cycle, replacement of axopodial arms, membrane and extrusomes, requires approximately 24 hours.

In Summary

Emphasis has been placed on phagocytosis in the naked amoebae and on ingestion by granuloreticulopodia (in foraminiferans) and by rhizopodial networks in the radiolarians. The co-operative effort in capturing active prey by pseudopodial arms in the heliozoans has been described and the probable function of extrusomes in supplying plasma membrane to the enlarged funnel pseudopod has been proposed.

The culture and biochemistry of *Entamoeba* has been described briefly because of its importance to the welfare of man. The ability to adopt facultative parasitism by *Naegleria* and *Acanthamoeba* allows these small amoebae to become human pathogens.

11 APICOMPLEXA

The Apicomplexa are obligate parasites which possess an apical complex consisting of one or more polar rings, a conoid (a spirally coiled structure inside the polar ring), a number of micronemes and a smaller number of rhoptries, elongated saccular organelles (Figure 11.1). The apical complex, usually present in the motile stage, is designed to allow the parasite to penetrate host cells. Many important genera are included in this phylum: *Plasmodium*, the causative organism of human malaria; *Eimeria*, a pathogen of domestic animals; *Toxoplasma*; *Sarcocystis*; *Isospora*; and the piroplasm *Babesia* which infects erythrocytes and causes red water fever in cattle.

Figure 11.1: A Generalized Diagram of the Apical Complex at the Anterior End of a Coccidian Merozoite, Showing Some of the Structures Associated with this Complex. co, conoid; im, inner pellicle membrane; mc, micronemes; om, outer pellicle membrane; pr, polar ring; rh, rhoptry.

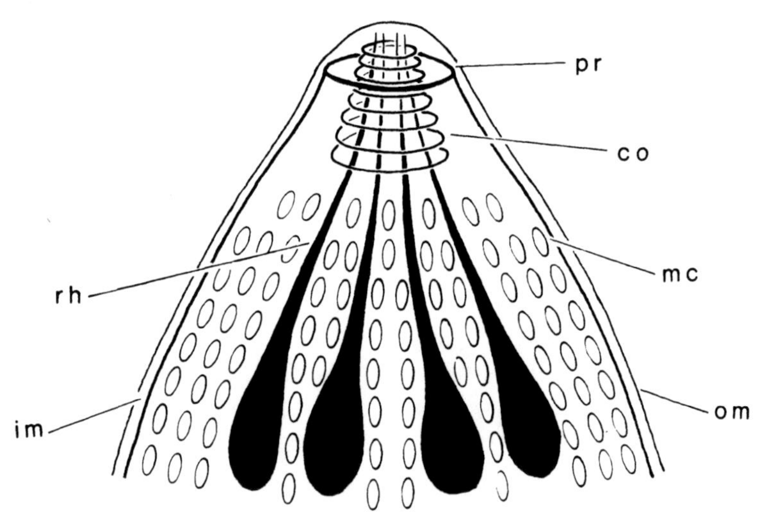

Source: Based on Hammond and Long (1973).

The classification being adopted here is that of Levine *et al.* (1980)

where the Apicomplexa are divided into two classes, the rather atypical Perkinsea and the major class, Sporozoea, with its three subclasses, Gregarinia, Coccidia and Piroplasmia. There is only one genus, *Perkinsus*, in the Perkinsea, which is separated from the other apicomplexans by two characters. The structure of the conoid and the production of zoospores which have flagella with mastigonemes are used by Levine (1978) as distinguishing class characteristics. *Perkinsus* owes its name to Perkins (1976), who found that this pathogen of the American oyster, *Crassostrea*, possessed an apical complex but with the conoid forming an incomplete ring structure.

Sporozoea

Within the Sporozoea interest has centred on the coccidians and the piroplams, relatively little having been published on the nutrition and biochemistry of the gregarines. The Gregarinia are extracellular parasites, mainly of invertebrates, so the apical complex is used as a semipermanent attachment organ such as is seen in the hooked epimerite of *Actinocephalus* or the mucron of *Gregarinia*.

Coccidians and piroplasms are intracellular parasites in their definitive hosts. A few are monoxenous, but the majority have an alternative host. The second host is normally a blood-sucking insect or an acarine (a tick). Patterns of life histories are too variable to present in a generalized scheme, therefore those which are of the most interest are described in the appropriate section.

Intracellular parasites must be able to recognize the appropriate cell type in the host, penetrate through the plasma membrane or stimulate invagination and, once inside, have a means of obtaining nutrients from the host cell cytoplasm. The presence of the parasite stimulates the host cell to produce nutrients, but only if it is the correct host cell. Sporozoites acquired accidentally have no difficulty in penetrating cells and surviving for a time, but unless the host is suitable no schizogony will take place and the parasite will disappear. The implication is that 'foreign' sporozoites, although physically able to enter cells, do not possess the appropriate genetic make-up to activate the cell.

The parasites, sporozoites and merozoites, have the specialized apical complex at their anterior end for penetration, although visual proof of its action is lacking at present. Conoids, paired organelles or rhoptries and possibly associated with them, polar rings, microtubules and micronemes, are implicated. The distribution and ultrastructure of these components of the apical complex of coccidians have been described in detail in Hammond and Long (1973).

The sporozoite or other infective body enters the host cell after making contact with the cell membrane. It may pass inside by puncturing the membrane which then closes behind it, leaving a narrow penetration line but more often the parasite enters by invagination. When this happens the parasite becomes enclosed in a section of the host cell membrane which forms the parasitophorous vacuole. In this vacuole the parasite feeds, grows and divides, protected from the host's antibodies. However, it must now obtain its food through two membranes, that of the parasitophorous vacuole and its own plasma membrane. Enough food must be absorbed to allow the parasite to make the appropriate transformation for passage to the next stage of the life cycle.

Patterns of life cycle vary from subclass to subclass and from genus to genus. The gregarinians have relatively simple monoxenous life cycles when the production of a protective cyst is essential for transmission of infection. The coccidians have several characteristic stages in their life history, transmit infection as a sporozoite that may or may not be enclosed in a spore. In species that pass through an insect vector, the need for a protective spore is superfluous. Phases of multiplication occur in both the asexual (schizogony) and sexual (sporogony) stages, interspersed with feeding and growth phases (Figure 3.10). The piroplasms are a small group of parasites of mammalian red blood cells or lymphoid cells. No spores are produced as transmission is direct using ticks as vectors. The piroplasm genera *Babesia* and *Theileria* reproduce only by asexual means (Figure 11.5).

Coccidia

The most useful and informative way to illustrate the Coccidia is to select a variety of examples and to use them to emphasize the most important characteristics. Assignment of species to the Coccidia is still rather uncertain, as is the misleading use of some generic names. The haemogregarines are heteroxenous members of the Coccidia in spite of their name suggesting affinity with the Gregarinia (Levine, 1982).

Haemogregarines live as parasites in the red blood cells of vertebrates, with sexual reproduction in an invertebrate. *Haemogregarina bigemina* is found in the marine blenny and is transmitted by the larva of the isopod *Gnathia maxillaris*, and possibly also by leeches (Davies, 1982).

The Eimerias

Plurality is a deliberate choice in this heading to indicate that the genus is a large one, including more than 95 species of *Eimeria*. In the Eimerias there is a high degree of host specificity; only closely related genera

Figure 11.2: The Life Cycle of *Eimeria tenella* in Domestic Fowl. Stages of asexual and sexual reproduction take place in the host gut epithelial cells; cysts are voided in the faeces and continue sporulation, ready to be eaten by a new host. Excystation and liberation of free sporozoites utilize the remaining amylopectin reserve as an energy source.

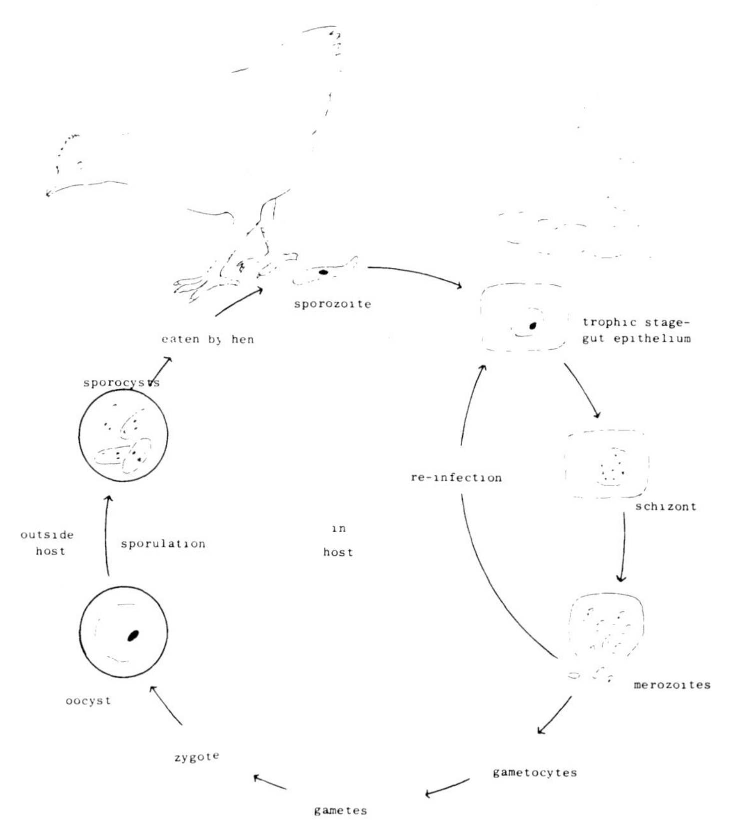

may become alternative hosts. Results of experimental cross-transmissions give rather conflicting results and perhaps should not be given too much credence. Indications are that domestic sheep and goats are alternative hosts, as are chinchillas, some wild rodents and laboratory rats and mice.

Infection starts when an infective sporozoite is released in the host gut and selects an appropriate epithelial cell to parasitize (Figure 11.2). It enters the cell by induced invagination, resulting in the parasite being enclosed in its parasitophorous vacuole of host cell plasmalemma. The absence of distinct feeding organelles and a cytostome are characteristic of *Eimeria*, but evidence is accumulating that there may be a site of ingestion of nutrients, the micropore. After invasion of the host cell, vacuoles containing degraded host cell cytoplasm appear first at the base of the micropore (a shallow depression with a shape characteristic of each genus) and then throughout the growing merozoite. Most species of *Eimeria* possess a single micropore located towards the anterior end of the organism. Characteristically it consists of the outer pellicle membrane invaginated into the shallow depression and lining it and the inner pellicle membrane invaginated far enough to form a double thickening in the neck but broken below. It seems certain that the micropore is the site of ingestion in the growing stages of the parasite. However, it has been found that pinocytosis, in association with V-shaped invaginations, is the means by which nutrients enter developing macro- and microgametocytes (Hammond *et al.*, 1967, 1969).

The nutrition of an intracellular parasite must inevitably be closely linked with that of the host cell, where the supply of nutrients should initially be ample. Growing schizonts of *E. tenella* require 14 vitamins. During early development of the schizont, the supply of nutrients will not be limiting and glycolysis is adequate as an energy-producing pathway. Much of the metabolic effort of the growing schizont is directed towards building up a reserve of carbohydrate. Amylopectin granules start to appear and increase until the onset of schizogony. The process of schizogony and the differentiation of new organelles use up most of the reserves of amylopectin in producing energy. However, continued growth results in gradual exhaustion of host nutrients and alternative pathways of metabolism may occur, a similar situation to that found in trypanosomes. During their brief extracellular period, the released merozoites show a transition to aerobic metabolism and the involvement of the Krebs cycle and a far more efficient use of substrates.

One important difference between *Eimeria* and trypanosomes is that the former relies heavily on its carbohydrate reserves to tide it over, especially in the oocyst stage, whereas trypanosomes die in the absence of exogenous substrates such as glucose. *Eimeria* reserve a carbohydrate which was originally thought to be glycogen (coccidien-glykogen). It has been shown to be amylopectin, which has been isolated and analyzed from *Eimeria tenella* and *E. brunetti* (Ryley *et al.*, 1969; Wang *et al.*, 1975). Granular deposits of amylopectin occur in the oocysts, sporozoites and merozoites in oval granules 0.5-1.0 μm in size, much larger than reserve granules of glycogen in metazoan cells (20-30 nm). The confirmation of granular amylopectin reserves in coccidians shows a remarkable parallel to the situation in rumen ciliate protozoa. This confirmation raises the question of whether or not the occurrence of amylopectin may prove to be more widespread in protozoa than was originally suspected. It may be that more rigorous analysis will show this to be true.

As released merozoites penetrate new gut epithelial cells some will transform into gamonts for the start of the sexual reproductive cycle. Young macrogamonts of those species examined are seen to accumulate amylopectin, but it is less common for microgamonts to behave similarly. *E. contorta*, from the intestinal epithelium of rats, is one species in which the presence of amylopectin in microgamonts has been confirmed (Mueller *et al.*, 1981). A developing gamont, consistent with its need to absorb nutrients efficiently, induces certain changes in the membrane of the parasitophorous vacuole. Numerous blebs and folds appear, projecting in towards the gamont, and probably providing an increased surface area for transport of host cell nutrients into the parasitophorous vacuole pool. Other structures characteristic at this stage are intravacuolar tubules which project from the surface of the developing macrogamont. It would be tempting to speculate that these tubules function for nutrient transport (uptake), but certain problems need to be explained first. These are that intravacuolar tubules appear to be empty of contents, which is surprising if they are used for nutrient transport, and they are absent in growing schizonts where the need for nutrients is as great, if not greater, than in gamonts. Further doubt is cast on the function of intravacuolar tubules due to their absence in developing macrogametes of *E. stiedai*. Pitillo *et al.* (1980) believe that the macrogametes of *E. stiedai* in the absence also of micropores, obtain their nutrients directly by the limiting membrane of the parasite and the host cell membrane of the parasitophorous vacuole being closely apposed to each other.

As gamonts develop, they become recognizable as macrogametocytes and microgametocytes. Although the ultrastructure of gametocytes has been investigated on many occasions there is no published research available on their metabolism. After fusion of gametes, oocysts develop

(Figure 11.2). There is no doubt that amylopectin is an energy reserve laid down for the oocyst and is to be consumed during sporulation, which is itself an aerobic process. Amylopectin reserves decrease from 83 to 46 μg per 10^6 oocysts during the first ten hours, then fall to 41 μg during the next five hours. After this there is a very slow resynthesis of amylopectin, as lipids have gradually replaced it as the principal source of energy, allowing conservation of reserves until a new host is found (Wilson and Fairbairn, 1961). The energy for excystation, release of sporozoites and host cell penetration is provided by the remaining amylopectin reserve.

Generally the protein requirements of coccidia are inadequately known. Synthesis of nucleic acids is very limited, indicating a reliance on degraded host cell nuclei as a source of nucleic acids.

In view of the importance of *Eimeria* as the causative organism of avian coccidiosis, possible lines of chemotherapy are being developed. Understanding of the more vulnerable enzyme systems provides targets for chemotherapy. One such enzyme is thymidylate synthetase which must be synthesized *de novo* by the parasite; it cannot be supplied by exogenous sources (Coles *et al.*, 1980). Without this enzyme the parasite would be unable to synthesize DNA as a result of thymine deficiency. This kind of approach is becoming increasingly important in treatment by chemotherapy.

Isospora and Sarcocystis

Isospora is a frequent and widespread parasite with many similarities to *Eimeria* in life cycle and in host specificity. *I. canis* in dogs infects epithelial cells of the intestine where the life cycle is completed to the oocyst stage. Other species of *Isospora* have been found with extra-intestinal stages. *I. serini*, one of a range of isosporas from passerine birds, develops in other tissues, lungs, liver and spleen, where it may form a reservoir for chronic infection of the host. *Isospora* closely resembles *Eimeria* in the morphology of the merozoite and gametocyte stages, but differences appear in sporulation. The oocyst of *Isospora* contains two sporocysts, each with four sporozoites, whilst *Eimeria* sporulates into four sporocysts containing two sporozoites – small differences perhaps in otherwise close genera.

Sarcocystis is a heteroxenous genus with specific definitive hosts, but variable intermediate hosts. Sporozoites emerge from accidentally swallowed oocysts and invade striated muscle tissue in various parts of the host body. Unlike *Eimeria* and *Isospora*, this sporozoite is not enclosed in a parasitophorous vacuole in the tissue cell. *S. cruzi*, parasitic in calves, is normally seen to enter a host cell by rupturing its limiting membrane. If it does enter by invagination then the host-enclosing membrane is soon lysed and disappears, leaving the parasite free in the cell. In this way the parasite is free from attack by host phagolysosomes. Schizogony leads to the production of large sarco-cysts in the host muscle.

Toxoplasma

This genus is represented by the single species, *Toxoplasma gondii*, which is related to *Eimeria* by life-cycle similarities, particularly in the oocyst stage. Its cosmopolitan nature and wide range of hosts make it a serious hazard. Successful completion of the life cycle requires two hosts, a cat and man or some other vertebrate. Gamogony leading to the production of oocysts takes place in the cat when immature oocysts are shed in the cat faeces. These sporulate outside the host to form two sporocysts of four sporozoites each. Sporulated oocysts are swallowed by the new host where the cyst wall ruptures releasing sporozoites into the gut (Figure 11.3).

The conditions necessary for rupture of the oocyst wall can be reproduced *in vitro* when excystment is induced by immersing oocysts in 5 per cent bovine bile in 0.9 per cent saline. After 30-60 minutes at 37°C the cyst wall ruptures. The released sporozoites pass through the gut epithelium and invade a variety of tissue cell types, macro-phages, nervous system, lymph nodes and liver, in man, birds, rodents and other higher mammals.

Asexual multiplication (schizogony) takes place in these host cells. In the host cell the growing first stage trophozoite has been given the name tachyzoite to describe its quickly multiplying behaviour. The parasite, enclosed in a parasitophorous vacuole, multiplies rapidly by asexual reproduction to produce little colonies of tachyzoites (Figure 11.3). The parasite is able to prevent host cell lysosomes from fusing with the parasitophorous vacuole and so destroying it (Jones, 1980; Jones and Hirsch, 1972). A parasitophorous vacuole at this stage has a distinctive appearance. The single membrane develops villous projec-tions and invaginations and the host cell endoplasmic reticulum and mitochondria attach to it giving the vacuole membrane a multi-layered appearance in places. These are some of the indications that invasion by the parasite increases the metabolic activity of the host cell (Khav-kin, 1981). Tachyzoites escape from the host cell by lysing the vacuole membrane, followed by rupture of the whole cell.

Tachyzoites may now pass into a second stage of slower multi-

Figure 11.3: The Life Cycle of *Toxoplasma gondii*. Stages of the life cycle in man, cat, and other vertebrates; with the sexual cycle in the gut epithelium of cats. Sporulated oocysts are swallowed by a vertebrate; infection multiplies rapidly as tachyzoites (acute infection), then slowly as bradyzoites in zoitocyst (chronic infection). Zoitocysts return to cat (through infected meat) and the sexual cycle is completed.

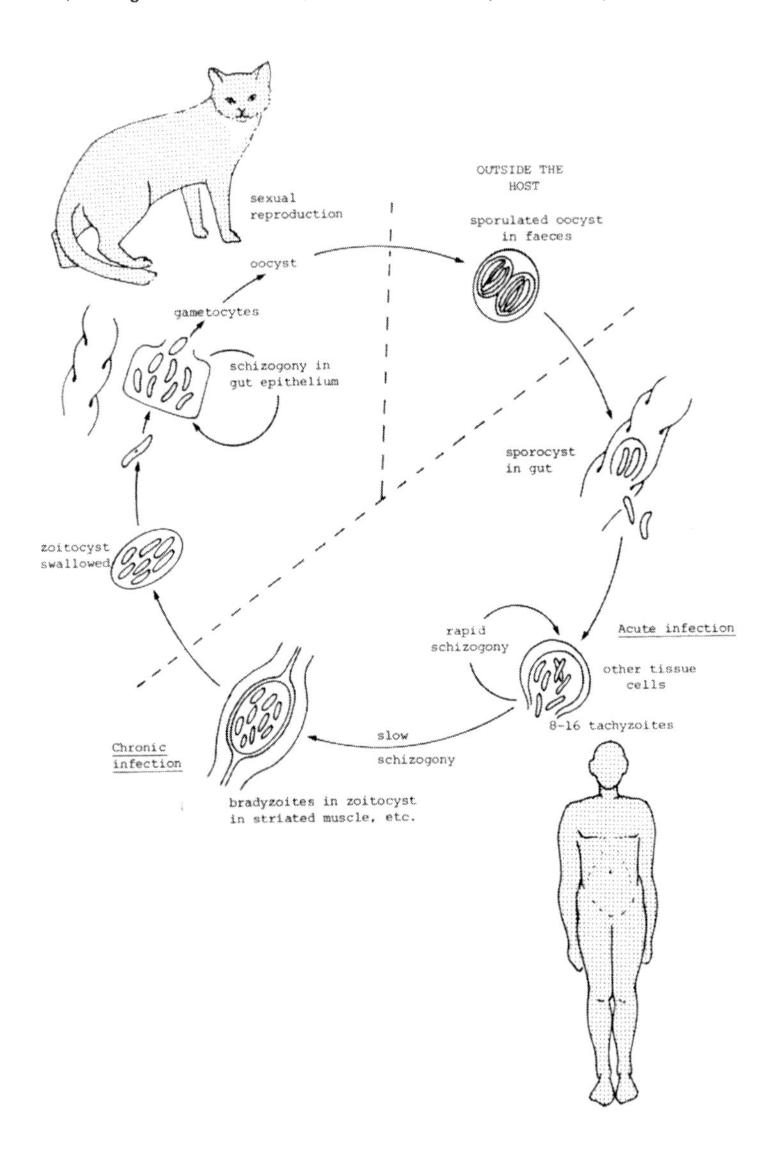

plication in cells of the heart, brain and muscle. These are the bradyzoites (slow organisms) which become enclosed in a zoitocyst, a tough long-lived structure which is responsible for maintaining chronic infections in the host. The zoitocyst remains an intracellular structure initially although bradyzoites may be released by rupture. Immunity from further infection is conferred on a host which has bradyzoites in its tissues but when this wanes, tachyzoites may develop again.

Infection passes back to a feline host when it eats the carcass of an infected animal. Zoitocysts release bradyzoites which develop in epithelial cells of the gut, growing into schizonts which multiply asexually or becoming gametocytes. This initiates the sexual cycle culminating in the production of infective oocysts which re-start the life cycle.

Most of the stages in *Toxoplasma* development are within host tissue cells which provide a steady source of nutrient for the growing parasite. In this situation very little reserve polysaccharide needs to be produced and the level of endogenous respiration is very low. Exogenous glucose is the preferred substrate for energy production and this is rarely in short supply. Although glucose is the favoured substrate, tachyzoites in *in vitro* experiments showed that glutamine supports active respiration with mannose, galactose, ribose, glucose-6-phosphate, pyruvate and fructose proportionately less effective. Respiration is normally aerobic and produces as end-products lactic acid, carbon dioxide and traces of volatile fatty acids (Fulton *et al.*, 1957, 1960).

The only stages when a store of energy-producing substrates is required are in the zoitocyst and oocyst stages. Energy for sporulation and release of the sporozoites until they establish in new host cells is provided by a reserve of amylopectin. Zoitocysts, which may have impaired nutrient uptake and may be long-lived, also rely on a reserve of amylopectin.

Although this description of *Toxoplasma* infections is very brief it has attempted to emphasize the points of interest in the biology of the parasite.

Plasmodium

Plasmodium is essentially a parasite of erythrocytes, although it spends part of its vertebrate stage in endothelial or liver cells and the sexual multiplication phase in an insect vector, usually a mosquito. A typical life cycle of *Plasmodium* is shown in Figure 3.10.

Erythrocytic Stage. The most active feeding stage in *Plasmodium* is that of the trophozoite growing in the red blood cells (the erythrocytic stage). The merozoite enters an erythrocyte by an invagination process which involves a sequence of clearly defined steps. This sequence has been described in *P. knowlesi*, a parasite of rhesus monkeys, and probably can be applied to most other species of *Plasmodium* in the erythrocytic stage (Johnson *et al.*, 1981). It involves recognition (of surface proteins), apical orientation, when the cone-shaped anterior end supported by rhoptries is pressed against the erythrocyte membrane, and release of rhoptry substance produced in the Golgi body. The merozoite then attaches to the erythrocyte membrane through specific receptor proteins on the merozoite surface. The superficial nature of these proteins is confirmed by the fact that they can be removed by trypsin treatment. After attachment the erythrocyte becomes deformed, bulging inwards as the parasite pushes into the invagination where it becomes enclosed in a membrane-lined parasitophorous vacuole. The motive force is generated by the parasite; the blood cell remains passive. Entry may be completed in some 10-20 seconds. Although the preceding description refers to *P. knowlesi* merozoites, sporozoites of *P. knowlesi* and *berghei* behave similarly.

The sporozoite is also seen to take an active part in penetration, but it relies on certain unidentified substances in normal serum before attachment can take place (Danforth *et al.*, 1980). Penetration is uninterrupted by treatment with Cytochalasin B, a phagocytosis inhibitor, which would block activity by the host cell. If the sporozoites are experimentally coated with antibodies, passage into the host cell does not protect them and they are destroyed.

Once inside the host cell a dedifferentiation process, shedding of the extrapellicular structures, takes place until the invader is only separated from host cell cytoplasm by two single membranes, one from the host cell and one from the parasite. It has become solely a feeding body with a definite site of ingestion, the cytostome, a feature now recognized to be common to most coccidians.

Plasmodium, as a parasite of erythrocytes, has available to it host cell cytoplasm, haemoglobin pigment and glucose in an aerobic environment. With a constant supply of glucose for energy production, there is no need of carbohydrate reserves, a similar situation to that in trypanosomatids. In the erythrocyte stage, *Plasmodium* trophozoites feed by phagocytosis, ingesting host cell cytoplasm through a small functional cytostome or micropore and into endocytotic vacuoles (Aikawa *et al.*, 1966). The actual process of ingestion is not fully understood, but

Figure 11.4: Ingestion Process in *Plasmodium* Trophozoite in a Host Erythrocyte. (a) The site of ingestion is a small cytostome (micropore) supported by two concentric rings round the opening; (b) host erythrocyte cytoplasm is drawn through the cytostomal opening into a swelling ingestion vacuole; (c) mature digestion vacuole. cy, cytostome; dv, digestion vacuole; e, erythrocyte; iv, ingestion vacuole, p, parasite.

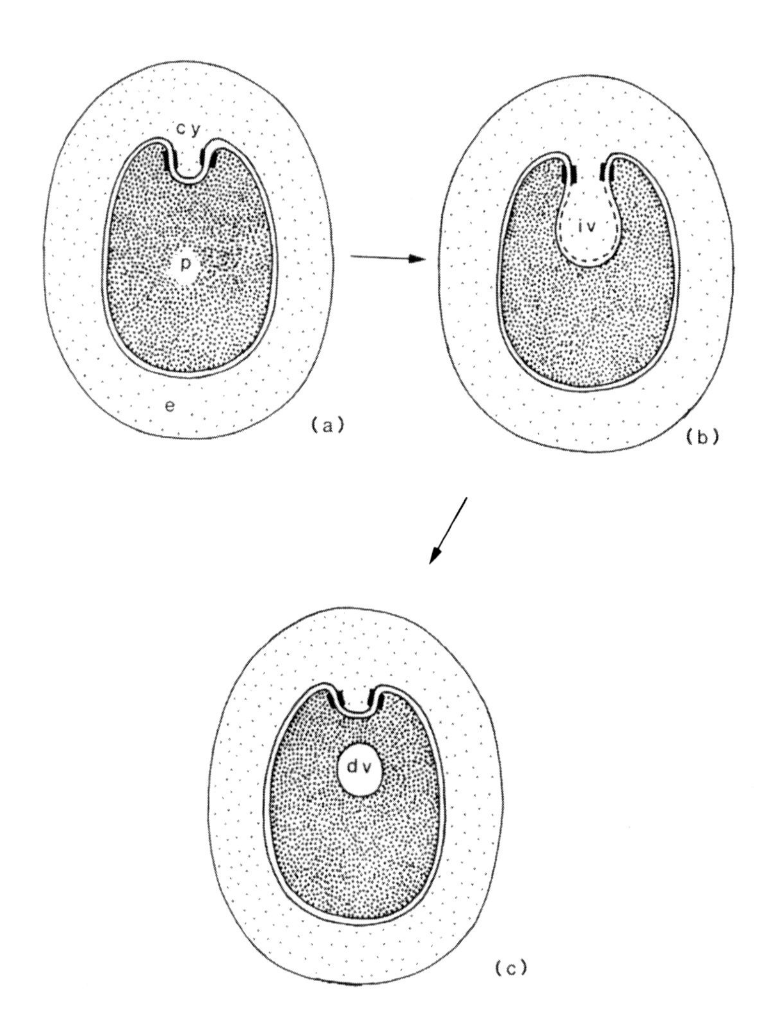

Source: Adapted from Aikawa (1971).

observations show that the shallow cavity of the cytostome bulges inwards, reaching up to 1 μm in length and taking host cell cytoplasm into the bulge. The size of the orifice of the cytostome remains unchanged throughout the process (Aikawa, 1971). The invagination continues to maximum size and then separates as an endocytotic vacuole enclosed initially in two membranes (Figure 11.4). As digestion commences, one membrane, that derived from the host cell, is lost due to the action of acid hydrolases.

Small variations in the ingestion organelle in different species of *Plasmodium* do not alter the general feeding pattern outlined above. *P. lophurae*, the causative organism of bird malaria, ingests host cell cytoplasm through a cytostomal invagination of the type just described, whereas the mammal species, *P. falciparum*, augments cytostomal ingestion with pinocytotic uptake through a series of small invaginations, resulting in many small endocytotic vacuoles. This is particularly apparent in the ring stage of trophozoite growth.

Digestion takes place within the single-membraned endocytotic vacuole where positive acid phosphatase activity is demonstrated histochemically. Following the initial hydrolysis of ingested food, metabolic pathways vary somewhat and this variation is reflected in the structure of the mitochondria and the activity of the Krebs cycle. Some species of *Plasmodium* have mitochondria which are distinctly cristate whilst in others they are acristate. *P. lophurae* of birds, with cristate mitochondria, possesses Krebs cycle enzymes and *P. knowlesi* of monkeys has acristate mitochondria and no Krebs cycle enzymes.

Haemoglobin, which is one of the principal nutrient sources, is only partially broken down in the process of extracting essential nutrients. The globin component is broken down to essential amino acids and the haem residue is converted to haemozoin for storage. It appears as dark pigment granules in the vacuoles of *P. lophurae* and *P. falciparum* (Deegan and Maegraith, 1956). The production of haemozoin as a waste (storage) product is important, for it results in the removal of free haematin from the system. Haemozoin is formed by the combination of haematin with a polypeptide. Free haematin would otherwise act as an inhibitor to succinic dehydrogenase, an important enzyme in the Krebs cycle for glucose oxidation (Sherman and Hull, 1960).

In *P. knowlesi* (of primates) and *P. berghei* (of rats), haemoglobin digestion takes place in small vesicles pinched off from the food vacuole to allow easy access of digestive enzymes. It is in these vesicles that haemozoin remains in granular form. Small vesicle digestion is necessary in these two species, as the food vacuole retains a double membrane. Haemoglobin from host cell erythrocytes provides the principal source of protein for the parasites. In *P. knowlesi*, 80 per cent of the methionine and cystine requirement is derived directly from the host cell, with the remaining 20 per cent from external (and unspecified) sources.

Although *Plasmodium* is fully equipped with all the enzymes necessary to metabolize glucose in a similar manner to the host cell, information on proteolytic enzymes suggests that these are different from the host erythrocyte. Two proteases have been isolated from *P. knowlesi* and *P. berghei*, one with optimum activity at pH 7.8 and the other at pH 4.5. The detailed biochemistry of *Plasmodium* has been presented fully elsewhere (Kreier, 1980).

Exoerythrocytic Stage. Feeding in other stages of *Plasmodium* is less clear. An operative cytostome, food vacuole formation and a residue of haemozoin granules have not been confirmed in the exoerythrocytic stage. Endothelial and liver cells, which house the exoerythrocytic parasites, have more heterogeneous contents and it may be that essential nutrients pass into the parasites by diffusion from the host cells.

Gametocytes. The cytostome, which has become operative in the erythrocytic stage (trophozoite), is retained in developing gametocytes which have a similar capacity to ingest host cell cytoplasm and to digest haemoglobin in endocytotic vacuoles. Aikawa *et al.* (1966) suggest that a special feeding organelle (the cytostome) has developed in the erythrocytic stage because larger quantities of haemoglobin are needed to supply essential nutrients.

Gametocytes pass into the mosquito gut to complete the sexual phase of the life cycle and produce infective sporozoites for passage back to a vertebrate host. Information on the insect phase has been slower to accumulate, partly due to the difficulty of developing *in vitro* culture methods. Much of the recent research on this stage has an immunological slant. For instance, immunizing the vertebrate host with *Plasmodium* gametes prevents fertilization of gametes in the mosquito gut, and consequent transmission of infection. The anti-gamete antibodies which are produced in the blood pass into the mosquito with the blood meal and cause surface membrane changes on the gametes. By treating *P. gallinaceum* gametes with mixed monoclonal antibodies raised against these gametes, Aikawa *et al.* (1981) were able to monitor gamete behaviour. Male gametes so treated develop an electron-dense surface coat and adhere in clusters lengthwise in long rope-like bundles. Female gametes react similarly in developing an all-over coating of antibodies and fertilization is blocked. This is one of the many approaches to the prevention of malaria transmission which may produce more promising results than chemotherapy.

Piroplasmia

The piroplasms are small heteroxenous blood parasites with patterns of life cycles similar to *Plasmodium*, which resulted in them being classified together initially. However, ultrastructural differences and the absence of haemozoin pigment in the piroplasms justify their separate subclass. The two principal genera are parasites of cattle, *Babesia* which multiplies in the red blood cells and *Theileria* which multiplies in lymphoid cells. More recently it has been recognized that *Babesia* can also infect human red blood cells with serious consequences in subjects who have had a splenectomy.

The life cycle of *Babesia bigemina*, an important pathogen in domestic cattle, is shown in Figure 11.5. Active binary schizogony spreads the parasite through erythrocytes, which are eventually destroyed. The parasite passes to *Boophilus*, a blood-sucking tick, invades its gut epithelial cells and increases asexually by schizogony. Vermicules are released which migrate directly to the ovaries or indirectly via the Malpighian tubules for further multiplication, before passing transovarially into developing ova. The ovum retains the parasite, which establishes first in the larval gut epithelium before migrating as a vermicule to the larval salivary gland. These infective salivary gland vermicules are injected into a new host when the larval tick attaches for blood-sucking.

The *Babesia* trophozoite is an actively feeding and growing parasite. For penetration into the erythrocyte *B. bigemina* has a rather different apical complex from that of the coccidians. The conoid part of the pellicle complex typical of coccidians is absent and is replaced by an 'Apikalschirm', an umbrella-like structure beneath the anterior pellicle. A ring of rhoptries lies below this umbrella.

In other species such as *B. microti* ingestion of nutrient is by non-cytostomal pinocytosis all over the surface, the parasite has no cytostome and does not phagocytose fragments of haemoglobin (Langreth, 1976). Other species may use cytostomal ingestion. Rudzinska (1976) has described a coiled organelle in *B. microti* which lies partly inside and partly outside the parasite. This she considers is responsible for

Figure 11.5: The Life Cycle of *Babesia bigemina* in Cow and Tick, *Boophilus annulatus*. The infection passes from the cow to the adult tick, to the larval tick (by transovarial transmission), then back to the cow.

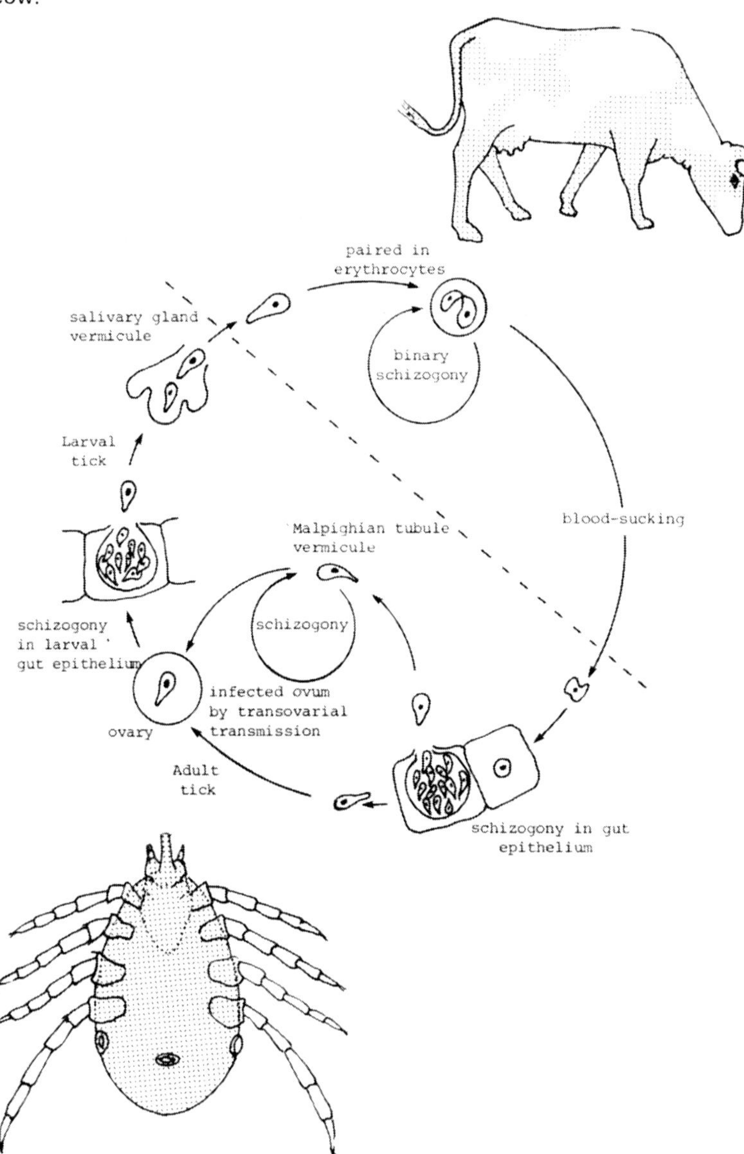

promoting extracellular digestion of host cytoplasm, providing soluble nutrients for pinocytotic uptake and aiding in the removal of waste products. The absence of haemozoin granules in *Babesia* suggests that it is capable of digesting haemoglobin more fully than is *Plasmodium*.

The present state of information on the biochemistry of these small piroplasmids is still rather scanty. A brief account of *Babesia* and *Theileria* may be found in Fulton (1969).

In Summary

This chapter contains only the briefest introduction to a large group of interesting and important parasites. An attempt has been made to draw attention towards the trends in current research thinking and to extract some of the more interesting findings.

Aspects of the biology of the coccidians, *Eimeria*, *Toxoplasma* and *Plasmodium* have been discussed. Recognition of the importance of piroplasms as human parasites as well as in cattle, has focused attention on them.

12 MYXOZOA, MICROSPORA AND ASCETOSPORA

The current classification revision requires that the Myxozoa and the Microspora be elevated to separate phyla on the grounds that they produce complex spores with extrusible polar structures, not found elsewhere in the protozoa. Previously the Myxozoa and the Microspora constituted a class of the protozoa, the Cnidospora. The term Cnidospora reflects a feature which they possess in common with the lower invertebrate phylum Cnidaria. On contact with a likely host the spore ejects a polar filament or tube which may aid in adhesion or penetration. A myxozoan spore, which is considered to be multicellular, ejects a filament for attaching the parasite to the host body. A microsporan differs in that it shoots out a tube which is used for injection of the infective organism into a host cell.

Myxozoa and Microspora contain genera which are of considerable economic importance. All are obligate parasites in a wide variety of animals which they infect by means of a small amoeboid sporoplasm that emerges from a characteristic spore. The myxozoans develop in body cavities and tissues, most commonly in cold-blooded vertebrates, where the sporoplasm grows into a large, multinucleate cyst-like growth while the microsporeans are intracellular parasites, especially of arthropods. It is appropriate to include the Ascetospora, formerly the Haplosporea, among these spore-producing parasites because, although their spores have neither polar filaments nor polar capsules, the general pattern of life history is similar to the microsporans. Whether the difference in spore structure is sufficient to justify a proposed separate phylum, the Ascetospora, remains for future debate. Information on all these groups is accumulating; new genera and species are being described and alternative hosts are being recognized, but identification can be difficult without much background information.

Spore Structure

The most readily identifiable stage is that of a mature spore whose structure is sufficiently distinctive to be used as a taxonomic aid.

Myxozoa. A typical myxozoan spore possesses polar filaments which

Figure 12.1: Mature Spores of Myxozoa and Microspora. (a) Myxozoan spore showing two polar capsules and bilateral symmetry typical of the family Myxobolidae; (b) microsporan spore showing a tubular filament which is not enclosed in a capsule, a polar cap and polaroplast. c, polar cap; pc, polar capsule; pf, polar filament; pp, polar plug; pt, polaroplast; s, sporoplasm; tf, tubular filament.

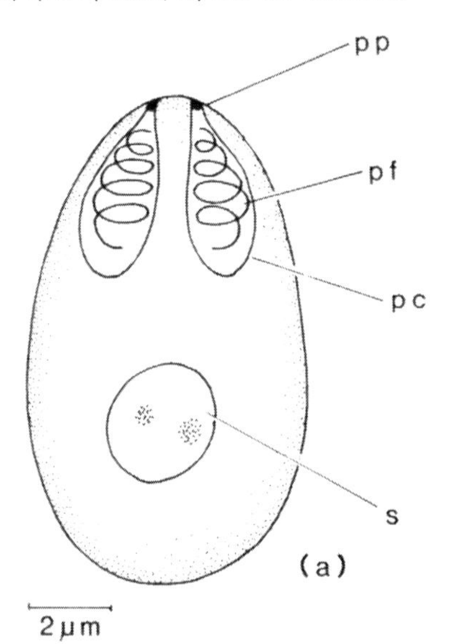

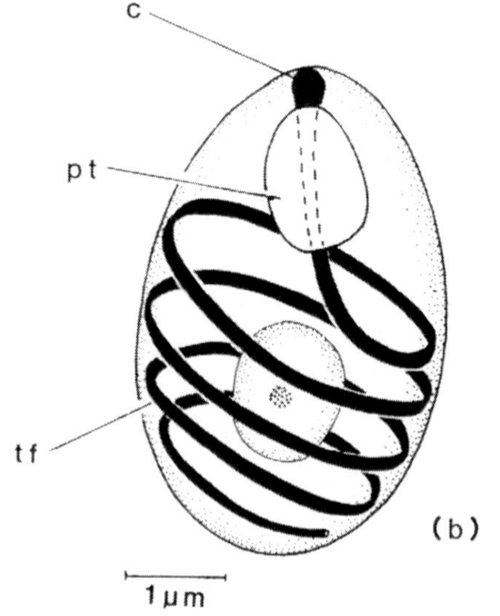

are enclosed in polar capsules (Figure 12.1(a)). The capsules are separated from the spore surface by a proteinaceous plug. The location and number of polar capsules varies from group to group; for instance the spore type illustrated in Figure 12.1(a) is typical of the Platysporina, a sub-order which includes *Myxobolus* and *Henneguya*. The bipolar arrangement found in the sub-order Bipolarina, which includes *Myxidium*, is illustrated in Figure 12.2(g).

Microspora. Although a microsporan spore possesses a prominent tubular filament, this is not enclosed in a capsule of the myxozoan pattern. The filament is a tube coiled within the spore and connecting to the surface through a polar cap whose polysaccharide nature is thought to be important in mediating ejection of the filament (Figure 12.1(b)). Another structure which is involved in the ejection of the tube is the polaroplast, a stack of smooth membranes continuous with the membranes of the tubular filament.

The manner in which the filament is ejected may be comparable with that in a nematocyst in *Hydra* where contact promotes the appropriate chemical action to increase the permeability of the polar cap or plug. Fluid enters, filling the polar capsule of the myxozoan or swelling the membranes of the polaroplast of the microsporan. The effect is to increase the internal hydrostatic pressure which forces the filament out, often in an explosive manner.

Myxozoa

The two classes Myxosporea and Actinosporea are very different in their representation and impact. Myxosporeans have attracted attention recently because of their influence on current commercial projects. The rapid increase in fish farming, particularly of Brown and Rainbow Trout, has created an ideal breeding ground for myxosporeans. The families Myxosomatidae (*Myxosoma*) and Myxobolidae (*Myxobolus* and *Henneguya*) are among the more serious hazards to fish farmers. *Myxosoma cerebralis* causes 'whirling' disease in salmonid fishes by extensive deformation of the cartilage of the head, axial skeleton and fins. *Myxosoma*, in common with all myxosporeans, produces spores which transmit infection to other host fish. Although one host is necessary for completion of the life cycle, it has been shown that birds play an indirect part in the spread of infection by eating dead infected fish. The spores pass through the bird alimentary canal unaltered, perhaps to be discharged in faeces in another body of water. A time of maturation in mud at the bottom of the pond seems to be a necessary part of the life cycle as spores shed into clean cement tanks are not readily infective.

The ecology of myxosporean infections is also interesting in that the site of infection in the host affects the way in which the parasite develops. The structure of the vegetative stage (the plasmodium) and of its outer layers, and its feeding behaviour as well as its pathogenicity, are all site-related. The more primitive parasites are all *coelozoic*, living in the fluids of the gall bladder, the urinary tract and the gill sinuses, and the more specialized are the *histozoic* parasites developing in the tissues of the gills, liver and muscles of the host. Some coelozoic and histozoic myxosporeans are listed in Table 12.1, with details of their host preferences.

Table 12.1: Some Common Myxosporea and their Hosts.

Parasite	Host
Coelozoic species:	
Myxidium gasterostei	Stickleback
M. macrocapsulare	?
M. serotinum	Salamander
Sphaeromyxa maiyai	Pacific tomcod
Chloromyxum catostomi	cyprinid fish
C. trijugum	sunfish
Zschokkella floridanae	killifish
Leptotheca minuta	rat-tailed fish
Histozoic species:	
Myxobolus spp.	various hosts
**Henneguya exilis*	Channel catfish (gills)
H. longicauda	" " "
**H. adiposa*	" " (adipose fin)
H. diversis	" " (liver, kidney)
H. pellis	Blue catfish (dermis)
Myxosoma cerebralis	salmonids (cartilage)
**M. funduli*	Atlantic minnow (gills)
Kudoa cerebralis	Striped bass (connective tissue of nerves)

Note: * Single-membraned plasmodia, atypical of histozoic species. See Figure 12.3(b).

Life Cycles

A typical life cycle of a myxosporean is seen in *Myxidium gasterostei*, a coelozoic parasite of the stickleback (Figure 12.2). The parasite

Figure 12.2: The Life Cycle of *Myxidium gasterostei* in the Stickleback. (a) Uninucleate amoeboid sporoplasm in the host; (b) multinucleate plasmodium; (c) formation of sporonts by compartmentalization of nuclei; (d) somatic nucleus secretes a spore envelope then disintegrates; (e) generative nucleus divides for the disporous condition; (f) each spore contains two nuclei to produce the polar capsules, two to produce the spore valves and two generative nuclei for the binucleate sporoplasm; (g) mature spore released from the host; (h) sporoplasm released by the explosive discharge of one polar filament and its two nuclei fuse.

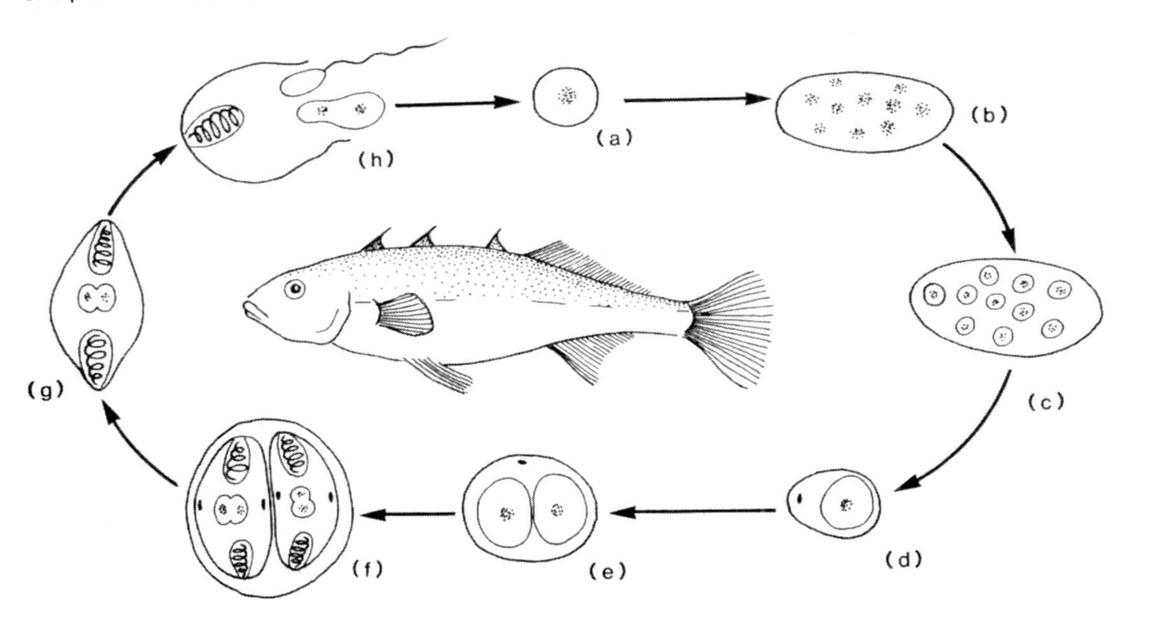

emerges from the spore as a binucleate sporoplasm in which the nuclei fuse to form an amoeboid infective body prior to rapid nuclear fission (a). The multinucleate plasmodium feeds and grows, finally changing character as many of the nuclei become surrounded by membranes and transform into sporonts (b-c). A series of nuclear divisions in the sporont follows (d-f), producing generative and somatic nuclei. At (d) the somatic nucleus secretes the spore envelope then itself disintegrates; the generative nucleus divides again (e), as *Myxidium* is a species which produces disporous sporonts or pansporoblasts. Further divisions (f) produce six generative and vegetative nuclei, of which two initiate the formation of the polar capsules, two valvogenous nuclei produce the spore valves and two form the binucleate sporoplasm. *Myxidium* spores are bipolar, having polar capsules arranged at opposite poles (g). When the sporoplasm is released (h) the two nuclei fuse, prompting the idea that they may be equivalent to gametes.

The need for the spore to remain in the water column for dispersion or at least in a position where it is most likely to gain entrance into a new host, is provided for by various flotation devices. Most spores are small and light with ridges and projections characteristic of the species. Spores of *Chloromyxum* show some interesting ecological adaptations in that *C. trijugum* parasitizing centrarchid fish (sunfish) of stiller waters has spore walls with three prominent ridges, whereas *C. catastomi* of cyprinid fish of streams and rivers has spores textured with elaborate ridges and pits to enhance flotation (Listebarger and Mitchell, 1980).

Host-parasite Relationships

The stage at which the parasite must exploit its host's resources to a maximum is the growing plasmodium, absorbing nourishment to prepare for sporogony and enlarging until it resembles a large, tumour-like cyst. Food is ingested through deeply-invaginated pinocytotic channels which develop in an organized layer in the plasmodium wall. Current, Janovy and Knight (1979) regard the plasmodium wall both as an important taxonomic feature and as an essential pinocytotic organelle for obtaining nourishment from the host. This view has been upheld by studies in several species of myxosporeans. Much of the detailed study on the plasmodium wall has been carried out on *Henneguya*, because of its economic implications. Depending on the site of infection (and the species) the plasmodium wall may be double-membraned or single-membraned. Typically the plasmodia of coelozoic species are enclosed in highly-convoluted, single-membraned walls with finger-like projections extending into the fluids of the gall bladder or urinary bladder of the host. A double limiting membrane usually develops round histozoic species.

Henneguya exilis, a pathogen in the gills of the Channel catfish, *Ictalurus punctatus*, develops different plasmodia according to whether the sporoplasm establishes itself within the gill lamellae adjacent to blood sinuses (intralamellar) or between the lamellae (interlamellar). These two types of plasmodia produce different clinical symptoms. Interlamellar plasmodia are much more damaging in that they cause extensive hyperplastic growth of the surrounding basal cells and fusion of the gill lamellae. These plasmodia are enclosed in double-membraned walls which are active sites of pinocytosis. Pinocytotic channels of two types are aligned at right angles to the surface of the plasmodium in a layer 0.7-0.9 μm thick. Single-membraned invaginations of the inner membrane only are not involved in ingestion; they are the 'stable' elements of the wall according to Current and Janovy (1976, 1978). Ingestion of host cell cytoplasm occurs at points where the plasmodium wall makes close contact with a host cell, when the outer membrane invaginates down into an inner channel to form a double-membraned channel (Figure 12.3(a)).

Clear evidence that this channelled layer plays an active part during the growth of the plasmodium is furnished by the presence of an underlying zone of well-developed mitochondria. In older plasmodia, once the major growth phase is completed and mature spores are present, pinocytotic channels, vesicles and mitochondria are much reduced. Greater metabolic potential is transferred to the maturing spores which must store up enough reserves for ex-sporulation once inside a new host. The sporoplasm when fully differentiated accumulates a reserve of β-glycogen particles and its cytoplasm contains mitochondria with prominent cristae, granular endoplasmic reticulum and free ribosomes, all evidence for the potential of efficient aerobic metabolism.

The alternative form of *H. exilis* plasmodia, the intralamellar plasmodia which grow adjacent to blood sinuses and those of *H. adiposa* which grow in the adipose fin of the Channel Catfish, behave rather as coelozoic species (Current, 1979). They are enclosed in a single-membraned wall with simple pinocytotic channels invaginating into the parasite ectoplasm (Figure 12.3(b)). Much less damage is caused to the host, as only interstitial material is ingested. A granular surface coat prevents direct contact with host cells.

Other myxozoans, the Actinosporea, parasitize oligochaete and

Figure 12.3: The Structure of the Pinocytotic Layer of *Henneguya exilis* Plasmodia in Different Sites in the Host. The nature of the contact between the parasite and the host tissue determines the extent of damage. (a) An interlamellar plasmodium with a double-membraned wall (dm) and two types of invaginated channels, pinocytotic channels (pc) and 'stable' channels (sc); (b) an intralamellar plasmodium with a single-membraned wall (sm) and simple pinocytotic channels (spc). pv, pinocytotic vesicles.

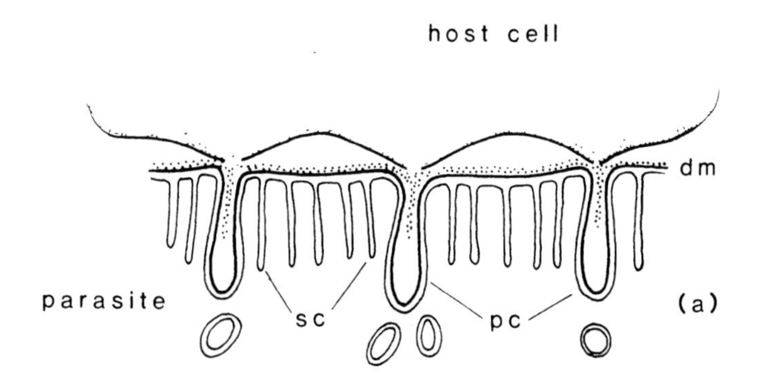

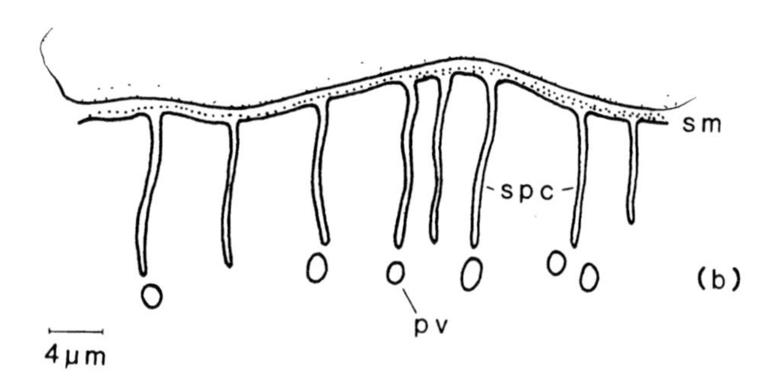

Source: Based on Current and Janovy (1978).

sipunculid worms and so have not attracted the same interest. They produce spores with three polar capsules and a membrane of three valves which may be smooth (*Sphaeractinomyxon*) or be drawn out into horn-like processes as in *Triactinomyxon*, a parasite of sipunculids.

Microspora

Microsporans are intracellular parasites which produce small unicellular spores, each containing a long coiled tubular filament and an infective body, a uninucleate or binucleate sporoplasm (Figure 12.1(b)). As the best-known genera belong to the order Microsporida the remainder of this section will concentrate on the microsporidans. Most parasitize arthropods: *Nosema bombycis* in silkworm moths, *N. apis* in honey bees and *N. algerae* and *Thelohania legeri* in anopheline mosquitoes where pathogenicity in the host may be of use in controlling insect-borne diseases. *Amblyospora*, which has been found in culicine mosquitoes, is transmitted transovarially, but is only fatal in males where it moves into adipose tissues and multiplies rapidly until the host is destroyed (Andreadis and Hall, 1979). Transovarial transmission of infection is also found in the non-pathogenic *Nosema parkeri* in ticks. Crustaceans provide hosts for *Ormieresia carcini* (in crabs) and *Thelohania* species (in shrimps), but some of the most interesting relationships are the many cases of hyperparasitism which occur between microsporidans and helminths, particularly in digenean larvae (Canning *et al.*, 1974; Canning, 1975). The list of species involved in this type of hyperparasitism includes five genera, *Nosema*, *Microsporidium*, *Pleistophora*, *Unikaryon* and *Stempellia*, with some fifteen species.

Glugea parasitizes fish where it forms xenomas or 'glugea-cysts' in distended cells of the intestinal wall and liver. *Glugea stephani* in pleuronectid flatfish is not usually fatal except where a heavy infection develops in young fish. Infection is most readily picked up in the upper estuaries, at temperatures above 15°C where the fish are feeding on small crustacea. It has been shown experimentally that heavier infections develop where spores pass through *Corophium* or *Artemia*, although whether this is because the crustaceans act as concentrators or because the spores are altered in some way is not clear.

Life Cycles

The development of microsporidans includes stages of multiplication, sporoblast production and maturation of spores which normally remain inside the host cell until the death of the host. *Nosema bombycis* was perhaps one of the earliest species to be investigated when its importance to the larvac and pupae of the silkworm moth was realized (Figure 12.4).

Figure 12.4: The Life Cycle of *Nosema bombycis* in the Larva and Pupa of the Silkworm Moth, *Bombycis*. (a) A binucleate sporoplasm released from the spore by extrusion of the tubular filament; (b) the sporoplasm penetrates an epithelial cell of the host gut; (c) schizogony to produce schizonts (for repeating the growth and division cycle) and sporonts (for spore production); (d) sporoblast; (e) a developing spore; (f) a mature spore released by rupture of the host cell.

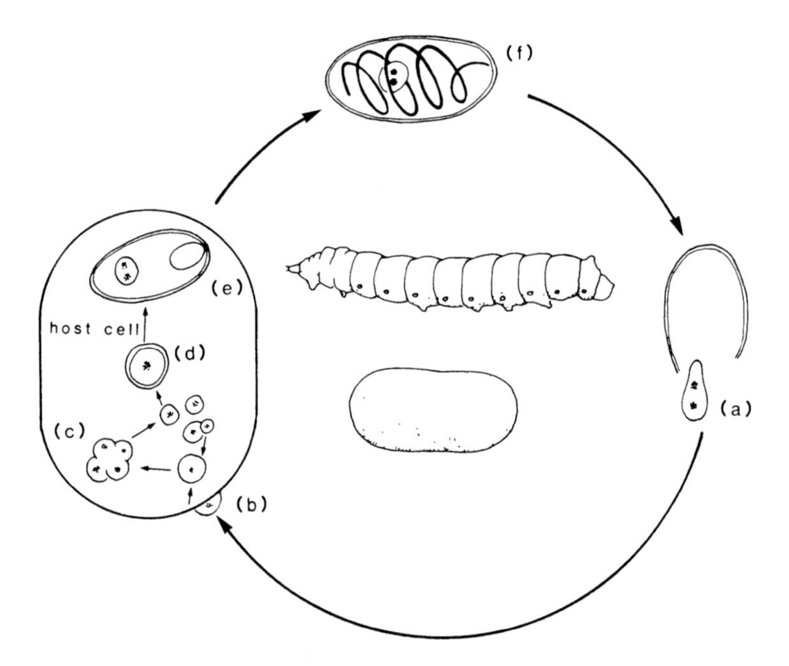

The binucleate sporoplasm is released from the spore when the tubular filament extrudes (a), and it penetrates a gut epithelial cell in which it will pass the major part of its development. At some point autogamous fusion of nuclei takes place (b). The intracellular body is enclosed only in a simple plasmalemma, so in effect it is in direct contact with the host cell cytoplasm, and yet cannot be attacked by host lysosomes. It absorbs nutrients through the plasmalemma. grows and divides by schizogony to produce schizonts and sporonts (c), which may either repeat the schizogony cycle or initiate sporogony (d-e). The exact relationship between these differentiating stages is not yet clear. The transformation of the sporont into a sporoblast involves thickening of the wall (d). Its further development includes the condensation of the tubular filament and polaroplast from vesicles of the Golgi system and the differentiation of a binucleate sporoplasm (e). On rupture of the host cell it is released as a mature spore (f).

Polysaccharides and acid phosphatases accumulate in the mature spore, particularly in the polar cap and around the tubular filament, both sites of great activity. Acid phosphatases located at this point in the spore must indicate some function relating to the extrusion of the filament and not to the normal digestive processes. Once inside the appropriate host the spore is activated through the polar cap and the tubular filament extruded. This filament is certainly implicated in the release of the sporoplasm, but whether as a tube down which the sporoplasm passes or simply by throwing it out when the tube everts has not been established in *Nosema*. In *Pleistophora* species nuclei are seen to pass down the tubular filament. How the infective amoeboid sporoplasm passes through the host cell membrane has not been explained with any certainty.

Host-parasite Relationships

Microsporidans are all intracellular parasites, so the host cytoplasm must provide all the necessary nutrients for a growth phase which commences with the entry of the minute sporoplasm and continues through vegetative reproduction to the stage at which the host cell is packed with mature spores. The large *Glugea* xenomas which develop in fish gut and liver tissues have provided some information and theories on food intake, although owing to their specialized structure it may not relate closely to other species.

A *Glugea* xenoma becomes an elaborate parasite-host complex, including much hypertrophying host tissue, measuring some 2-4 mm (Weidner, 1976). The xenoma is enclosed in a plasma membrane with an overlying mucopolysaccharide surface coat. This membrane profile consists of tubular processes which are thought to be stable elements and membrane processes which are transient elements whose staining properties suggest that they are responsible for secreting the glycoprotein component of the surface coat. The tubular processes which increase the surface area of the xenoma and the surface coat together are responsible for the active transport of metabolites into the growing parasite. There is no evidence of pinocytotic uptake of host cytoplasm comparable with myxosporean plasmodia. Weidner (1976) believes that developing schizonts live in harmony with host cytoplasm and obtain their energy from host mitochondria as they accumulate in the peripheral zone where host mitochondria are most numerous. It is only during sporogenesis and maturation of spores, towards the centre of the xenoma, that extensive destruction of host cell organelles becomes obvious.

Production of a host-parasite complex also occurs in *Ichthyosporidium giganteum*, another parasite of fish (Sprague and Vernick, 1974). The infection develops in subcutaneous connective tissue of the abdominal wall as a large syncytial xenoma, within which parasite and host tissue exist in physiological unity. This extreme hypertrophy of host tissue appears to serve as a protection to the parasite, creating a barrier between it and possible attack from host toxic or antigenic substances.

There are still many gaps in the story of host-parasite relationships in the microsporidans. New techniques are being developed for culturing these parasites by artificially infecting mammalian cells in tissue culture, and this step should help towards establishing the pattern of nuclear events, the means by which the parasite obtains its nutrients and what nutrients it requires. *Nosema algerae* and *N. eurytremae*, both parasites of a wide range of hosts, including arthropods and other invertebrates and some vertebrates, can be induced to infect monolayer cultures of pig kidney cells and rat brain cells especially when the spores are centrifuged on to the monolayers. Centrifugation seems to increase greatly the chances of infection. Whether this is simply due to improved cell-spore contact, whether it forces spores into the cells or whether it induces germination in some way is still a matter for speculation (Smith and Sinden, 1980).

Ascetospora

The Ascetospora transmit infection through the agency of simple spores which do not possess any form of extrusion apparatus. However, their general pattern of life history is close enough to the Microspora for them to be included here. *Minchina* and *Marteilia* have attracted attention by being parasites of oysters since these bivalves are farmed commercially. Although they normally occur in the digestive gland of the oyster, they are rather erratic in their ability to infect a host. The American native oyster, *Crassostrea virginica*, is becoming resistant to infections of *Minchina nelsoni* and it is only in introduced stocks of oysters that high mortality occurs (Farley, 1975). Other members of the Ascetospora (*Haplosporidium*) are parasites of marine annelids of no commercial importance with the result that less research effort is directed towards them.

In Summary

This chapter has reviewed briefly three lesser known groups of parasites, Myxozoa, Microspora and Ascetospora. The myxozoans produce distinctive spores with polar capsules containing coiled filaments. These parasites develop in the body cavities and tissues of commercially farmed fish and other hosts. Microsporans are intracellular parasites which infect a wide variety of hosts through spores containing an infective sporoplasm and a simple extrusion apparatus. Precise information on food intake and nutritional requirements is not very far advanced.

13 INTER-RELATIONSHIPS IN PROTOZOAN COMMUNITIES

In the first few chapters a picture was built up of different kinds of habitats and the trophic levels which exist in them. Following this the protozoan phyla have been considered in a systematic way to illustrate how different organisms trap and utilize the food they require.

Food supply and competition for food are the most important factors in determining whether a particular protozoan community can develop, providing certain physical and chemical limits are not exceeded. In a body of water protozoa will select their habitat according to the distribution of their preferred food supply. As the majority of protozoa take relatively passive food, bacteria and algae, swimming about in the water column will not prove a very profitable activity.

Benthic marine foraminiferans illustrate clearly the principle of distribution according to preferred food supply as most are highly selective feeders. There are filter-feeders, browsers and a few carnivores. The majority browse on diatoms, which means that they are restricted to the photic zone inshore and that the type of shore is sheltered with a stable substrate of fine sand or silt or low rocks on which an algal film can develop. Foraminiferans which are bacterivores can live in deeper waters, making use of symbiotic associations with zooxanthellae to supplement or substitute for food in nutrient-poor areas (Lee, 1980b).

Competition for Food

Competition for food, when certain items are in short supply, operates at many levels. That those protozoa which collect food more efficiently and multiply quickly will thrive at the expense of other protozoa which perhaps take longer to complete a division cycle, is one view. These are the successful pioneer species which show r-selection (Henebry and Cairns, 1980a). Yet this kind of reasoning must be used with caution, for it does not separate biomass of protozoa from their growth rate. Luckinbill (1979) takes an alternative view by considering that the best competitors are those which have a slower growth rate but achieve a larger biomass, in other words the K-selected species. Although *Para-*

mecium primaurelia grows more slowly it has a larger biomass and lies first in Luckinbill's (1979) rank order: *Paramecium primaurelia* > *P. tetraurelia* > *Colpidium striatum* > *Tetrahymena pyriformis*. These organisms are all competing for the bacterium *Enterobacter aerogenes* in his tests, and are not subjected to predator attacks.

Another form of competition which results in food-sharing has been described in the foraminiferan species *Allogromia*. *Allogrômia laticollaris* thrives in a culture of abundant algae (*Dunaliella* and *Nitzschia*) and low bacteria numbers, for it takes algae preferentially (Lee *et al*., 1966a). *Allogromia* sp. (unidentified) feeds as well on bacteria as on algae. When the two *Allogromia* species are cultured together *Allogromia* sp. takes the bacteria and leaves the algae for *A. laticollaris*, unless a disproportionately large population of *Allogromia* sp. is introduced. In this situation the bacterial numbers may be inadequate to support *Allogromia* sp., then it starts to feed on algae, bringing both species of *Allogromia* into direct competition.

Density of Food

There is plenty of evidence that the density of the food supply affects the efficiency of feeding and, consequently, the growth rate of protozoa, from *Tetrahymena pyriformis* (Curds and Cockburn, 1971; Villarreal *et al*., 1977); *Colpidium campylum* (Dive, 1975; Laybourn and Stewart, 1975; Taylor, 1978); and *Uronema nigricans* (Berk *et al*., 1976). *Uronema marinum,* when feeding on *Vibrio* sp. in chemostat culture, showed that cell size is directly related to the density of the prey (Ashby, 1976). The extreme example of the effect of the density of the prey is when the prey is so scarce that feeding becomes unprofitable and energy is consumed in collecting food and maintaining general metabolism faster than it can be replaced.

Phagotrophic protozoa of the open sea can satisfy their feeding requirements by browsing over dead copepods or aggregations of floating detritus where bacteria have collected. The density of bacteria elsewhere is not enough. A marine species of *Uronema*, feeding on bacterium *Serratia marinorubra*, needs 3.2×10^6 bacteria per ml when it can double its population every 4.6 hours (Hamilton and Preslan, 1970). An even shorter generation time in *U. marinum* (2.5 hours) allows this species to exploit suitable micro-environments such as the bacterial growth on decaying littoral matter (Parker, 1976).

Data from the *in vitro* culture of the foraminiferan *Allogromia*

Figure 13.1: The Relationship between Ciliate Growth and the Concentration of Suspended Bacteria.

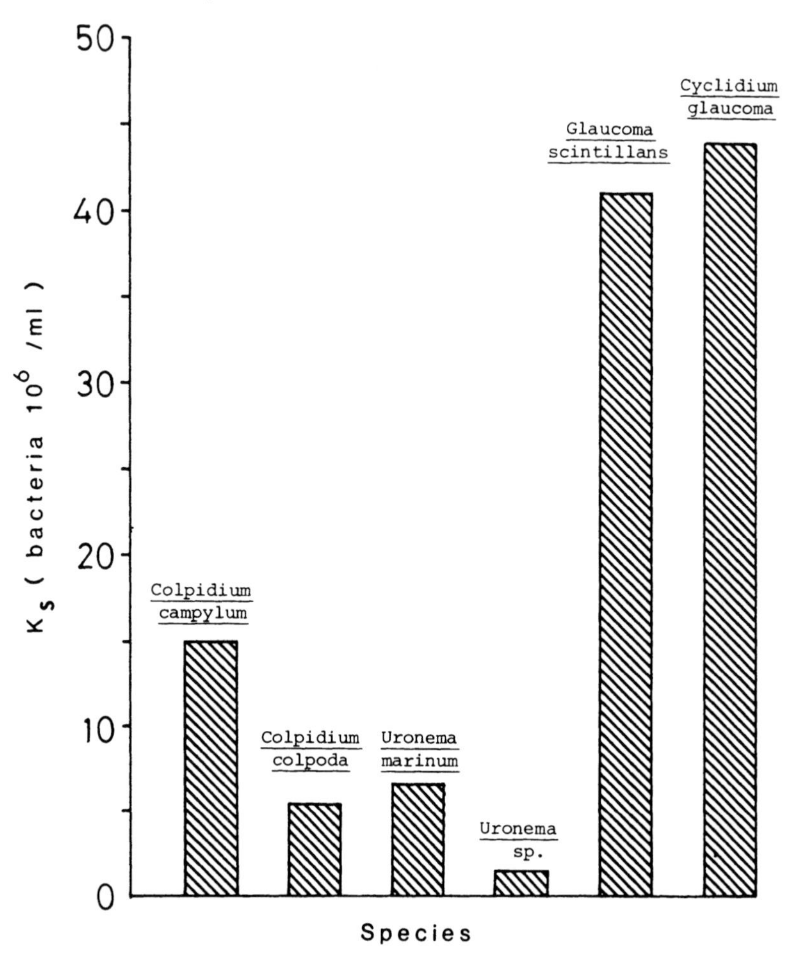

Source: Ashby (1976), Hamilton and Preslan (1970) and Taylor (1978).

laticollaris, mentioned above, shows that this organism begins to feed very erratically if the density of its algal food falls below 10^3 organisms per 10 ml culture. An ideal concentration in the culture vessel is in the range 10^3-10^5 algal cells per 10 ml culture (Lee *et al*., 1966a).

Erratic feeding and growth rate are not necessarily due entirely to low density of food organisms, for accessibility is also an important factor. A ciliate such as *Glaucoma scintillans*, which is an obligate deposit feeder, grows erratically when its bacterial food supply is kept in suspension by gentle shaking (Taylor, 1978). Yet when the bacteria are allowed to settle, *Glaucoma* is able to graze on the settled bio-film, and this grazing habit enables it to grow on a lower density of settled bacteria than a suspension-feeding ciliate can. Again this is an example of a protozoon exploiting the growth style of its prey. The suspension-feeding ciliate cannot easily feed on settled bacteria, for its oral membranelles are not adapted for grazing (Fenchel, 1980d).

Confirmation that keeping bacteria in suspension definitely favours suspension-feeding ciliates is shown in Figure 13.1. The suspension-feeders *Colpidium* and *Uronema* grow effectively on a much lower concentration of bacteria than do *Glaucoma* and *Cyclidium*, as shown by plotting growth rate against the concentration of suspended bacteria and calculating K_S values (half saturation values) of the curves.

Predator-prey Relationships

In a predator-prey relationship where the predator is a hunter, the density of the prey has a profound effect on the feeding efficiency of the hunter. *Didinium* feeding on *Paramecium* encounters its prey by random collision. Therefore when preying on a high-density population of *Paramecium* in a fairly homogeneous environment, it could quickly wipe out the paramecia, whereas in a low-density population hunting would be so unrewarding that *Didinium* might die of starvation. This type of interaction between *Didinium* and *Paramecium* has been described by Luckinbill (1973, 1974), Maly (1978) and Salt (1979).

Various devices for artificially impairing the hunting efficiency of *Didinium* have been tested: increasing the viscosity of the medium, providing starved prey (Luckinbill, 1973; Salt, 1979) and reducing the collision rate by lowering the density of the prey (Luckinbill, 1974; Salt, 1979). Some apparently conflicting results emerge for which there is no obvious explanation.

Providing very low-density, half-starved prey has a stabilizing influence on the population because *Didinium* swims more slowly, reducing its chances of random collision (Luckinbill, 1973). By slightly increasing the density of starved prey (to 3 organisms per μl), *Didinium* swims abnormally quickly, which has a destabilizing effect on the culture (Salt, 1979). The adaptive value of abnormally fast swimming when faced with scarce starved prey is clear. The predator is able to seek out new areas and more plentiful prey. An alternative strategy is for part of the population to encyst, reducing the number of active predators until a new prey population re-establishes (Salt, 1967).

The preceding sections have considered interactions between organisms in relatively confined communities, mostly under experimental conditions. Interactions on a larger scale and involving more mixed communities occur in all habitats, but the more extreme the habitat, the more restricted and specialized will be the distribution of organisms in that community. Two such habitats have been selected, the very hot and the very cold, to illustrate the effects of extreme physical and chemical conditions on the composition of communities.

Communities in Thermal Areas

Areas of geothermal activity, particularly those in North America, Iceland and New Zealand represent special ecological environments in which species diversity is limited by extreme physical and chemical conditions. Ground water wells out of hot springs at near boiling point, but this seemingly hostile environment supports actively-growing thermophilic bacteria. Very hot water drains into outflow channels and as it cools a clearly-defined temperature gradient develops over a distance which relates to the size of the spring. The importance of being able to carry out ecological studies along a single outflow channel with graded temperatures, but with chemical constituents remaining relatively constant, was recognized by Brock (1978).

Hot springs are of two types, the acid sulphate and the alkaline chloride. Acid sulphate springs are semi-stagnant mud pools which release pockets of hydrogen sulphide gas and few organisms will tolerate this combination of very hot acidic water and sulphurous gas. Alkaline chloride springs, even at temperatures of 99-101°C, have been shown to support life, the aerobic bacterium *Thermus aquaticus*, which is not merely existing but is growing vigorously. There is evidence that anaerobic bacteria have been isolated from such springs. Along with a slightly alkaline pH, these hot springs contain high concentrations of chlorides and silicates, the latter particularly coming out of solution as the water cools and forming the coloured sinter terraces which attract the eye of a visitor. The hot water which bubbles out of the ground is never totally devoid of oxygen, but this remains at a very low level in the outflow channel until the water cools to at least 35°C.

It seems that prokaryotes can tolerate higher temperatures than eukaryotes, as the next life forms to appear are blue-green bacteria (cyanobacteria). *Mastigocladus laminosus* and species of *Synechococcus*, both typical thermophilic cyanobacteria, begin to consolidate mat-like growths at 70-73°C. *Oscillatoria* and *Phormidium* appear later. The eukaryote alga *Cyanidium caldarum* normally develops vigorous growth at 40°C, although it can exist at 60°C in acid springs in the absence of the more vigorous cyanobacteria.

Figure 13.2: Hot Spring Effects, Waimangu, New Zealand. A V-shaped mat of cyanobacteria and algae develops as the temperature falls. Growth at sites 1 and 2 is *Mastigocladus* (cyanobacteria); at site 3, mixed but mainly *Spirogyra* (Chlorophyta).

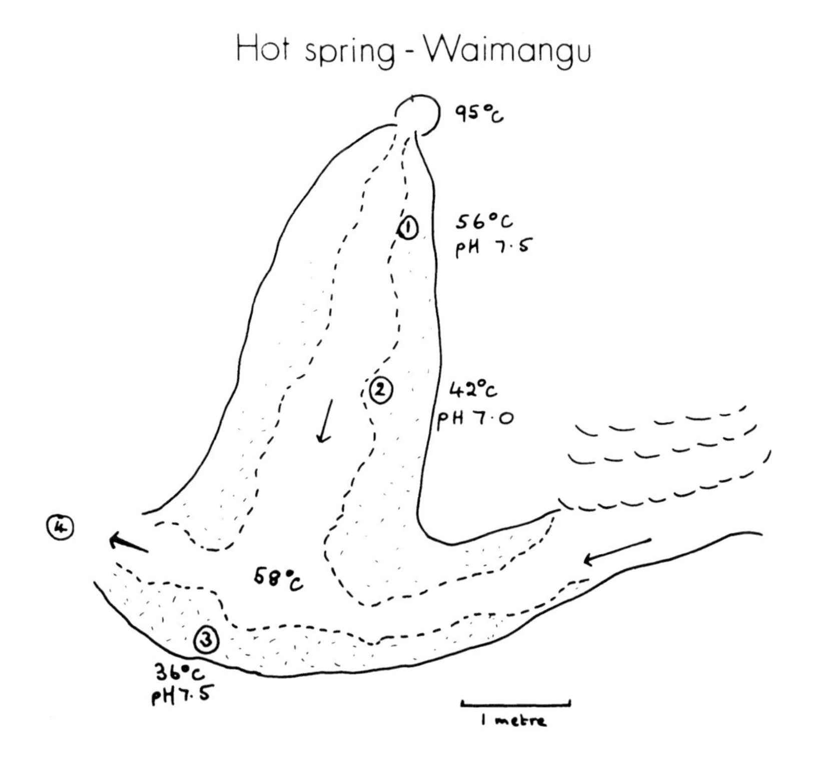

Cyanobacterial and algal mats form the basis of the hot spring communities. As the water fans out from the spring and cools, organisms grow at the cooler margins of the flow to give a V-shaped pattern of

Table 13.1: Upper Temperature Limits for Protozoa in Thermal Areas.

	°C	Source
Mastigophora:		
Chilomonas paramecium	36	Nisbet, 1978*
Sarcodina:		
Actinophrys sol	38	Issel, 1906
Amoeba proteus	45	Issel, 1906
A. verrucosa	40	Uyemura, 1936
Centropyxis sp.	45	
Difflugia sp.	45	W & B, 1967†
Hartmanella sp.	54	Hindle, 1932
Hyalodiscus limax	54	Issel, 1910
Quadrula sp.	45	
Trinema sp.	45	
Vahlkampfia limax	51	Uyemura, 1936
Ciliophora:		
Aspidisca lynceus	36	Nisbet, 1978*
Chilodonella cucullus	40	Nisbet, 1978*
C. algivora	45	Nisbet, 1978*
Chlamydodon cyclops	45	Nisbet, 1978*
Colpidium colpoda	33	W & B, 1967†
C. campylum	40	Nisbet, 1978*
Colpoda cucullus	39	Issel, 1906
Ctedoctema acanthocrypta	45	W & B, 1967†
Cyclidium citrullus	58	Kahan, 1972
C. glaucoma	51	Issel, 1910
Enchelys sp.	43	Issel, 1906
Euplotes patella	35	Issel, 1906
Frontonia acuminata	50	Issel, 1906
Glaucoma scintillans	44	W & B, 1967†
Halteria grandinella	36	Nisbet, 1978*
Histriculus vorax	40	Nisbet, 1978*
Holostichia vernalis	33	W & B, 1967†
H. hymenophora	33	W & B, 1967†
Kahlia acrobates	33	W & B, 1967†
Lionotus fasciola	40	Uyemura, 1936
Loxocephalus granulosus	36	Issel, 1906
Metopus sigmoides	45	Issel, 1906
Microthorax pusillus	36	Nisbet, 1978*
Nassula elegans	50	Issel, 1910
Oxytricha fallax	45	Issel, 1906
Oxytricha sp.	45	Nisbet, 1978*
Paramecium aurelia	42	Nisbet, 1978*
P. calkinsi	45	W & B, 1967†
P. caudatum	40	Uyemura, 1936
Tachysoma pellionella	45	Nisbet, 1978*
Urocentrum turbo	36	Nisbet, 1978*
Uroleptus limnesi	36	Nisbet, 1978*
Urostylia sp.	35	W & B, 1967†
Vorticella sp.	45	Nisbet, 1978*

Notes: † W & B, Winterbourn and Brown (1967)
* Nisbet (1978) is unpublished material.

growth (Figure 13.2). Once the rather glutinous algal mat has established, this forms a substrate on which other bacteria and detritus collect, providing a varied diet for browsing animals. Although the small bacterivorous ciliate, *Cyclidium citrullus*, has been found at 58°C by Kahan (1972), there is no certainty that it was able to divide at this temperature. In laboratory cultures he found that it would not divide at temperatures above 47°C. This illustrates one of the problems of studying the distribution of protozoa in hot springs. Are these species growing and dividing or simply surviving? There is no real evidence of any obligate thermophilic protozoa but Table 13.1 lists the highest temperatures at which different species have been found. Apart from the isolated example of *Cyclidium citrullus*, small amoebae seem to tolerate the higher temperatures. *Naegleria fowleri*, the causative organism of fatal amoebic meningoencephalitis in humans, which has established itself in some hot pools, has already been described in Chapter 10.

In general, protozoa resemble other animals in temperature tolerance, having an upper temperature limit of 46-47°C. Even at this temperature very few are deemed to be active; they may be encysted and only activate on cooling when brought back to the laboratory. Temperature tolerance can be monitored over short distances in the field. Figure 13.2 shows the changes in temperature from a hot spring until the water joins the main channel and also illustrates the extent of micro-organism growth along the margins. No protozoa are found at Site 1 (56°C), five species at Site 2 (42°C), and ten species at Site 3 (36°C). The distribution of species recorded is listed in Table 13.2. Temperature must remain one of the dominant factors in controlling protozoan distribution, but other factors should also be considered.

The types of micro-organisms which settle and grow to form a consolidated substrate are clearly important. Cyanobacteria in themselves are not a favoured food source for many protozoa, but how many other types of bacteria and how much detritus accumulates in the substrate does affect colonization by protozoa. A fresh growth of *Mastigocladus* near the water source supports a poorer population of protozoa than a softer mat of *Oscillatoria* further down the water flow, even if the temperature is similar. The implication is that the mobile filamentous *Oscillatoria* traps potential food for protozoa more efficiently than leathery mats of *Mastigocladus*.

This brief look at thermal areas as habitats for protozoa poses many questions to which there are no certain answers at present. There is no general pattern of food preference, mostly a 50-50 mixture of bacteri-

Table 13.2: Establishment of a Protozoan Population in a Hot Spring Run-off, Waimangu, New Zealand.

Source spring: 95 °C, pH 7.5.

Temperature °C	56	42	36
Metres from source	1	3	7
Substrate growth	*Mastigocladus*	*Mastigocladus*	*Spirogyra*
Ciliata			
Cyclidium glaucoma		+	+
Paramecium aurelia		+	+
Chlamydodon cyclops		+	
Urocentrum turbo			+
Microthorax pusillus			+
Vorticella sp.		+	
Halteria grandinella			+
Oxytricha sp.		+	+
Aspidisca lynceus			+
Uroleptus limnesi			+
Sarcodina			
Actinophrys sol			+
Mastigophora			
Chilomonas paramecium			+
Total species	0	5	10

Source: Nisbet (unpublished)

vores and herbivores. All can tolerate high salinities (having contractile vacuoles), temperatures up to the mid-forties, and often very low concentrations of dissolved oxygen.

Communities in Cold Climates

Much of the information on the types of habitats available to protozoa in extreme cold climates has come as a result of the establishment of scientific bases on the islands of the Antarctic. Two major types of habitat are under investigation, the land masses mainland and islands, and the surrounding ocean. The maritime land masses warm up enough in the summer months to allow growth of micro-organisms similar to cold temperate regions. In consequence, the records of protozoa from these regions contain no great surprises, a large population of testate amoebae, a few naked amoebae, and a diverse population of ciliates and flagellates, typical of wet terrestrial habitats. There is no evidence of geographical barriers to the spread of protozoan populations, provided they can produce small resistant cysts. To survive they must have the capacity to encyst during the winter months, more as a protection against drought than cold, for in winter ground water is frozen. Vegetation type and its underlying layer of compact fibrous peat is the most important factor in protozoan distribution. The rate of decomposition is very slow, so that organic material accumulates and very acid conditions prevail. Smith (1978) recorded 90-98 per cent organic content and pHs of 3.0-4.5 in peat below moss-turf in Antarctica.

Many of the Antarctic islands have a ground cover of mosses (*Polytrichum*, *Brachythecium*) or the antarctic grass, *Deschampsia antarctica*, all of which contribute to the underlying layer of peat and produce conditions similar to acid moorlands in northern temperate regions, including northern England. Sandon and Cutler (1924) and Flint and Stout (1960) found that antarctic soils have a flora and fauna comparable with temperate soil. Smith (1978) recorded terrestrial protozoa from 97 sites on Antarctic and sub-Antarctic islands and found a species composition of 124 species as follows: 31 flagellates, 10 naked amoebae, 35 testate amoebae and 48 ciliates. Of these he considered that 83 species could justifiably be counted as established members of the antarctic fauna. His study of habitats and associated communities of protozoa is a major work and makes interesting reading. Some 21 species, common temperate ones, including the ciliates *Colpoda cucullus* and *C. steini*, are not found south of 60°S and are probably limited rather by the angiosperm vegetation than by climate. Distinctive alkaline habitats produced by animal guano are included in his survey, where he found coprozoic and polysaprobic species.

The small cosmopolitan testacean, *Corythion dubium*, apparently thrives in Antarctic grassland peats where it was found to be the dominant species forming up to 80 per cent of many populations (Heal 1965). Smith (1973a) found *Corythion* equally ubiquitous in the moss-turf peats of Signy Island down to a depth of 9 cm, with a clear spring bloom following increases in bacteria and yeasts as the ground warmed up.

In many areas very thick layers of peat accumulate below moss or other vegetation because the rate of microbial decomposition is slow, during only the short summer season. This is no indication of reduced numbers of micro-organisms, for in fact the microbial population of soils from Signy Island was found to have a higher population than a moorland of the north of England (Heal *et al.*, 1967). This comparison is shown in Table 13.3

Table 13.3: Micro-organisms in Soils of Signy Island and an English Moorland.

	Micro-organisms/cm^3 Soil	
	Signy Island	English Moorland
Bacteria	7-10 x 10^9	2·8 x 10^9
Testate amoebae	9 x 10^3	7 x 10^3

Source: Heal *et al.* (1967).

The climate does not affect the numbers and range of species of bacteria and fungi so much as soil type and vegetation. The species which live in these regions are cosmopolitan, cold-tolerant species, although once established they may become further cold-adapted according to Heal *et al.* (1967). Similar conditions prevail in the Arctic soils of Spitzbergen (Sandon, 1924) and in east Greenland where Stout (1970) recorded a range of ciliates and testate amoebae consistent with cool temperate regions, in comparable habitats.

Bacterial populations correspond closely to those of temperate regions, but another potential food source for protozoa, yeasts, develop much higher numbers than in temperate regions (Baker, 1970), perhaps responding to high acidity. They appear to show no seasonality. However, yeasts may be of limited value as food for protozoa, as their numbers fall off markedly with depth, whilst bacteria increase rapidly (Table 13.4).

Reasons for this switch of emphasis in micro-organisms can be conjectured and probably relate to acid, aerobic conditions at the peat surface favouring yeasts. Following down through the 12 cm profile, decreasing oxygen and moisture, coupled with less extreme acid conditions, favour bacteria preferentially.

This switch in microbial distribution is reflected in the vertical zonation of testate amoebae for whom soil bacteria must form the major food source in the deeper layers. Apart from *Corythion dubium*, which shows no vertical zonation through 0-9 cm, the number of species of testaceans generally increases with depth to 10 cm in *Sphagnum*.

Table 13.4: Microbial Numbers at Different Depths from Peats of Signy Island.

Depth (cm)	Colonies x 10^3/gm Wet Wgt Soil	
	Bacteria	Yeasts
1-2	88	520
6-7	173	204
11-12	302	10

Source: Baker (1970).

However, it is dangerous to jump to the conclusion that vertical zonation is controlled by bacterial food supply, oxygen and moisture. Many testate amoebae which live among sphagnum contain symbiotic algae, zoochlorellae, which means that they are effectively contained in their distribution to the upper parts of the moss growth where light is adequate. *Hyalosphaenia papilio* and *Amphitrema flavum*, both zoochlorellae-bearing species, flourish in the top 6 cm; below this they become pale (Heal, 1962).

Another factor which has been shown to affect vertical zonation of testaceans is the availability of suitable particulate material for test-building, peat and mineral particles and diatom frustules in particular. As particulate matter is washed down to accumulate in the lower levels of the moss carpet, it is here that larger numbers of species of testates are found. Heal (1962) found the following distribution of species in sphagnum plants which he divided into sections: upper section, 8 species of testate amoebae; middle section, 12 species; lower section, 20 species. There was some overlap in those species which had to compromise between the light requirements of their zoochlorellae and test-building materials.

Not surprisingly the influence of animal faeces affects the microbiology of certain areas. Deposits of guano related to penguin rookeries produce a habitat which is rich in nitrogen, but because of the nature of bird faeces, more alkaline than the naturally vegetated sites. Smith (1973b) identified three species of ciliates, *Tetramitus rostratus*, *Philaster armata* (?) and *Vorticella microstoma*, which characteristically appeared in association with the alga *Prasiola crispa*, which colonized the nitrogen-rich ground after penguins left the rookery. *Vorticella microstoma* is a polysaprobic indicator, a bacteria-feeder which is also very common in sewage effluents. Testaceans were absent from this habitat.

The Polar Oceans

Marine protozoa of the polar oceans are mainly planktonic ciliates of the open water, the polyhymenophoran tintinnids (Littlepage, 1968), or browsing ciliates which spend part of the year under the sea ice (Fenchel and Lee, 1972).

Below the hard layer of sea ice is a layer of 'slush', loosely aggregated crystals which harbour a community of interstitial micro-organisms which play a significant part in the productivity of polar seas. The main agents of primary productivity in this under-ice layer are diatoms, which apparently receive light through the ice to maintain a rate of photosynthesis which does not fall far short of that for planktonic algae in open water (Bunt, 1963). In early January samples of seepage water from the slush layer below 4 metres of hard ice in McMurdo Sound, show maximum numbers of diatoms, 16.8×10^6 per litre. Of the 32 species of diatoms found, most form the epontic community, either attached to or closely associated with, ice crystals. At times of maximum productivity of the epontic community, planktonic diatoms below the ice are in such low numbers that they cannot easily be quantified. Light is being cut out from further penetration by the density of the diatom layer under the ice (Bunt and Wood, 1963). Associated with the algal layer is a wide variety of interstitial protozoa, mostly temperature-tolerant species. However, Fenchel and Lee (1972) identified a cold-water fauna, five ciliate species, as being physiologically adapted to these conditions, including two new species.

Considering the rather limited range of food available to protozoa living below the sea ice, it is not surprising to find that Fenchel and Lee's cold-water fauna, species of *Uronema*, *Holosticha* and *Uronychia*, feed on unicellular algae (diatoms), and that the new species described, *Spiroprorodon glacialis* and *Euplotes antarcticus*, are predaceous on other protozoa in the sea-ice interstices. Nor is it surprising that most of the ciliates found in this interstitial habitat are crawling species or, at least, ones which associate with surfaces in some way.

In areas where the sea ice breaks up in the Antarctic summer, the integrity of this interstitial community of micro-organisms is lost, although productivity of planktonic species will be enhanced. In McMurdo Sound where the ice layer may remain intact, primary productivity, followed by protozoan growth, will be very slow to build up in the Antarctic spring. As Bunt (1963) found, diatom populations do not reach a maximum until early January.

The two types of environment discussed, the very hot and the very cold, show extreme contrasts, particularly in physical conditions. However, they do not show a large range of protozoan species which are restricted only to these environments. Provided that appropriate food supply is present, many protozoa can adapt to extreme conditions.

Joint Enterprises

Many joint enterprises are set up in the protozoan world. Perhaps the closest type of relationship is that which exists between protozoa and their endosymbionts, so close that *Paramecium bursaria* can digest its symbiotic *Chlorella* in times of crisis – not the fairest kind of relationship.

Cases of endosymbiosis are frequently recognized and described, although not so well-understood is the actual nature of the symbiotic association. The behaviour of the dinoflagellate zooxanthellae which live in the gelatinous envelope of a colonial radiolarian *Collozoum* is typical of photoautotrophic organisms. They show diurnal rhythm, congregating near the surface of the colony or spreading through the rhizopodial network during daylight and at night they are concentrated deeper round the host cells (Anderson, 1980). *Collozoum* digests its symbionts and in conditions of normal illumination these zooxanthellae can maintain their numbers by asexual multiplication. In enforced darkness the number of symbionts dwindles. The concept that the migrations of zooxanthellae and the rate at which they are digested is part of a cultural enterprise by the host radiolarian, is mentioned in Chapter 10. A similar inter-relationship exists in the foraminiferans and their symbionts (Lee, 1980b).

Ciliates establish themselves in the ruminant stomach working hard in producing carbohydrates and proteins for the ruminant and in return receiving enough to grow and reproduce their own kind. The approach to understanding this relationship is two-fold: to isolate and culture the ciliates *in vitro*, examining their nutritional intake and output, and to study the performance of the host ruminant in the presence and absence of ciliates. Metabolic studies on these ciliates are described in Chapter 8.

Hypermastigid flagellates digest cellulose and other plant materials in the hindgut of woodroaches and termites, thereby forming an obligate symbiotic association. Each is dependent on the other. Outside the host these anaerobic flagellates die unless they are enclosed in protective cysts. Joint enterprises in this group extend beyond metabolic activities. The reproductive behaviour of the flagellates is linked to the growth cycle of the host woodroach. Living in the hindgut, where the exoskeleton lining is shed at every moult, the flagellates respond to the host moulting hormone by enclosing themselves in protective cysts. Outside their host they can reproduce sexually and survive in the cyst until eaten by a new insect host.

Termites and their flagellates have a different strategy which does not involve cysts, but which seems equally successful. Protozoa shed at the moult die outside the host. Newly-moulted termites acquire a new population of symbionts by inducing non-moulting insects to excrete moist faeces with active protozoa and these are eaten directly. The apparent lack of sexual reproduction in termite flagellates and the fact that they respond to the host's moulting hormones by dying, seems to indicate that they owe their success to the chance that termites practise coprophagy.

Tropic Levels in Free-living Communities

The essential interdependence of units in a community is as real for micro-organisms as it is for larger animals. This line of thinking has been presented in some of the early chapters. The dependence of protozoa on algae and bacteria in fresh waters, in marine environments and in the soil has been stressed. Very thriving micro-communities develop until the protozoa increase faster than the food source can be replenished, resulting in some action by the protozoa, either moving on to new pastures or encysting, otherwise perishing.

This burst of activity is particularly obvious in the summer months when a combination of increased light and warmth boosts the populations of algae and bacteria (Figure 2.5). A summer group of ciliates can be identified which flourish for a time and then apparently disappear. They select from the food organisms available. *Paramecium caudatum* feeds on vigorous growths of bacteria, *Loxodes striatus* takes benthic diatoms and *Frontonia leucas* takes cyanobacteria (Finlay *et al*., 1979).

Population explosions are not necessarily restricted to the summer months. The autumn overturn in a eutrophic pond may release trapped nutrients and allow a late bloom of algae, which is followed by a sudden increase in large benthic ciliates. *Loxodes magnus* appears at such times, feeding on the alga *Scenedesmus* (Goulder, 1972). Here *Loxodes* makes no real impact on its food supply, algae are always in excess, and maximum feeding takes only 0.68 per cent of the standing crop of *Scenedesmus*. Temperature, not food supply, will be the limiting factor on this population growth.

These are a selection from the many studies on trophic levels in free-living communities. Reference to others may be found in Chapters 2 and 3.

In Conclusion

Protozoa have very limited powers of movement in terms of foraging for food. Having to rely on cilia or flagella and certainly on pseudopodia will not move the organisms over long distances. Some have tried to solve their transport problems by attaching to larger animals and living as ectocommensals. This solution appears to work especially among peritrichs, chonotrichs and others which have developed efficient mechanisms for trapping food as it passes. The disadvantage is that they no longer have a free choice where they go, unless they detach and take a chance in finding a new host.

Protozoa in standing water lead a relatively static existence, with limited migrations in lateral and vertical directions and exploiting local food sources as they occur. Marine protozoa make full use of currents and tides for transport and for renewing supplies of nutrients. Their problems are not the result of consuming all the food available but of preventing themselves from being washed away by the tide. They respond to tidal rhythms and have developed a body shape which allows them to migrate down through the substrate as the tide recedes.

In sewage treatment plants protozoa face rather similar environmental conditions to marine protozoa. Their food supply is virtually continuous, but to survive they must either attach themselves to the surfaces of stones (in percolating filters) or to flocs of bacteria (in activated sludge), or be flattened efficient 'creepers' which can avoid being washed out.

Parasitic protozoa have been considered, not in terms of joint enterprises, but in relation to how they can make the best use of the food available to them in their host's body and how they can make provision for the continuance of their line.

BIBLIOGRAPHY

Aaronson, S. (1980) 'Descriptive biochemistry and physiology of the Chrysophyceae' in M. Levandowsky and S.H. Hutner (eds.), *Biochemistry and Physiology of Protozoa*, vol. 3 (Academic Press, New York), pp. 157-8

—and Baker, H. (1961) 'Lipid and sterol content of some protozoa', *Journal of Protozoology*, *8*, 274-7

Abou Akkadar, A.R. and Howard, B.H. (1960) 'The biochemistry of rumen protozoa 3. The carbohydrate metabolism of *Entodinium*', *Biochemical Journal*, *76*, 445-51

—and Howard, B.H. (1962) 'The biochemistry of rumen protozoa 5. Nitrogen metabolism of *Entodinium*', *Biochemical Journal*, *82*, 313-20

Adam, K.M.G. (1959) 'The growth of *Acanthamoeba* sp. in a chemically defined medium', *Journal of General Microbiology*, *21*, 519-29

Agersborg, H.P.K. and Hatfield, W.D. (1929) 'The biology of a sewage treatment plant', *Sewage Works Journal*, *1*, 411-24

Aikawa, M. (1971) 'The fine structure of malaria parasites', *Experimental Parasitology*, *30*, 284-320

—Hepler, P.K., Huff, C.G. and Sprinz, H. (1966) 'The feeding mechanism of avian malarial parasites', *Journal of Cell Biology*, *28*, 355-73

—Rener, J., Carter, R. and Miller, L.H. (1981) 'An electron microscopical study of the interaction of monoclonal antibodies with gametes of the malarial parasite, *Plasmodium gallinaceum*', *Journal of Protozoology*, *28*, 383-8

Aleem, A.A. (1949) 'The diatom community inhabiting the rural flats at Whitstable', *New Phytologist*, *49*, 174-88

Allen, R.D. (1974) 'Food vacuole membrane growth with microtubule-associated membrane transport in *Paramecium*', *Journal of Cell Biology*, *63*, 904-22

—(1978) 'Membranes of ciliates: ultrastructure, biochemistry and fusion' in G. Poste and G.L. Nicholson (eds.), *Membrane Fusion* (Elsevier/North Holland, Amsterdam), pp. 657-763

—and Staehelin, L.A. (1981) 'Digestive system membranes: freeze-fracture evidence for differentiation and flow in *Paramecium*', *Journal of Cell Biology*, *89*, 9-20

Anderson, O.R. (1978) 'The feeding behaviour, nutrition and reproduction of *Thalassicolla nucleata*', *Tissue and Cell*, *10*, 401-12

— (1980) 'Radiolaria' in M. Levandowsky and S.H. Hutner (eds.), *Biochemistry and Physiology of Protozoa*, vol. 3 (Academic Press, New York), pp. 1-42

Andreadis, T.G. and Hall, D.W. (1979) 'Development, ultrastructure and mode of transmission of *Amblyospora* sp. (Microspora) in the mosquito', *Journal of Protozoology*, *26*, 444-52

Andren, L., Lindgren, P.E., Nyholm, K.G. and Olsson, J. (1968) *Marinbiologi*, Götesborgs Vattendvården läggningar Undersökningar för havsutsläpp år 1966-67, Huvudrapport

Antia, N.J. (1980) 'Nutritional physiology and biochemistry of marine cryptomonads and chrysomonads' in M. Levandowsky and S.H. Hutner (eds), *Bio-*

chemistry and Physiology of Protozoa, vol. 3 (Academic Press, New York), pp. 67-115

Antipa, G.A. (1977) 'Use of commensal protozoa as biological indicators of water quality and pollution', *Transactions of American Microscopical Society*, *96*, 482-9

Ardern, E. and Lockett, W.T. (1928) *Manchester Rivers Department Annual Report*, Appendix 1, 41-6

Arme, C. (1982) 'Nutrition' in F.E.G. Cox (ed.), *Modern Parasitology* (Blackwell, Oxford), pp. 154-9

Ashby, R.E. (1976) 'Long-term variations in a protozoan chemostat culture', *Journal of Experimental Marine Biology and Ecology*, *24*, 227-35

Bader, F.G., Tsuchiya, H.M. and Frederickson, A.G. (1976) 'Grazing of ciliates on blue-green algae: effects of ciliate encystment and related phenomena', *Biotechnology and Bioengineering*, *18*, 311-31

Bailey, R.W. and Gaillard, B.D.E. (1965) 'Carbohydrases of the rumen ciliate *Epidinium ecaudatum*', *Biochemical Journal*, *95*, 758-66

—and Howard, B.H. (1963) 'Carbohydrases of the rumen ciliate *Epidinium ecaudatum* 2. α-galactosidase and isomaltase', *Biochemical Journal*, *87*, 146-51

Baker, J.H. (1970) 'Quantitative study of yeasts and bacteria in a Signy Island peat', *British Antarctic Survey Bulletin*, *no. 23*, 51-5

Baker, W.R. and Buetow, D.E. (1976) 'Hydrolytic enzymes of *Euglena gracilis*: characterization as a function of culture age and carbon deprivation', *Journal of Protozoology*, *23*, 167-76

Bamforth, S.S. (1971) 'The numbers and proportions of testacea and ciliates in litters and soils', *Journal of Protozoology*, *18*, 24-8

—(1980) 'Terrestrial protozoa', *Journal of Protozoology*, *27*, 33-6

Band, R.N. and Cirrito, H. (1979) 'Growth response of axenic *Entamoeba histolytica* to hydrogen, carbon dioxide and oxygen', *Journal of Protozoology*, *26*, 282-6

Bardele, C.F. (1972) 'A microtubule model for ingestion and transport in the suctorian tentacle', *Zeitschrift für Zellforschung und mikroskopische Anatomie*, *126*, 116-34

—(1974) 'Transport of materials in the suctorian tentacle' in M.A. Sleigh and D.H. Jennings (eds.), *Transport at the Cellular Level*, Society of Experimental Biology Symposium, no. 28, Cambridge University Press, Cambridge, pp. 191-208

—and Grell, K.G. (1967) 'Electronen mikroskopische Beobachtungen zur Nahrungs aufnahme bei dem Suktor *Acineta tuberosa* Ehr.' *Zeitschrift für Zellforschung und mikroskopische Anatomie*, *80*, 108-23

Barker, A.N. (1942) 'Seasonal incidence and occurrence of sewage protozoa', *Annals of Applied Biology*, *29*, 23-33

—(1943) 'The protozoan fauna of sewage disposal plants', *Naturalist*, Hull, July-Sept., 65-9

Barsdate, R.J., Fenchel, T. and Prentki, R.T. (1974) 'Phosphorus cycle of model ecosystems: significance of decomposer food chains and effect of bacterial grazers', *Oikos*, *25*, 239-51

Bassham, J.A. and Calvin, M. (1957) *The Path of Carbon in Photosynthesis* (Prentice-Hall, Englewood Cliffs, N.J.)

Bauchop, T. (1980) 'Scanning-electron microscopy in the study of microbial digestion of plant fragments' in D.C. Ellwood, J.N. Hedger, M.J. Latham,

J.M. Lynch and J.H. Slater (eds.), *Contemporary Microbial Ecology* (Academic Press, London), pp. 305-26

Berens, R.L. and Marr, J.J. (1979) 'Growth of *Leishmania donovani*: amastigotes in a continuous macrophage-like cell culture', *Journal of Protozoology*, *26*, 453-6

Berk, S.G., Colwell, R.R. and Small, E.B. (1976) 'A study of feeding responses to bacterial prey by estuarine ciliates', *Transactions of American Microscopical Society*, *95*, 514-20

Bick, H. (1972) *Ciliated Protozoa*, an illustrated guide to the species used as biological indicators in freshwater biology (World Health Organization, Geneva)

—(1973) 'Population dynamics of protozoa, associated with the decay of organic materials in freshwater', *American Zoologist*, *13*, 149-60

Bisalputra, T. (1974) 'Plastids' in W.D.P. Steward (ed.), *Algal Physiology and Biochemistry* (Blackwell, Oxford), pp. 140-2

Blum, J.J. and Rothstein, T.L. (1975) 'Lysosomes in *Tetrahymena*' in J.T. Dingle and R.T. Deane (eds.), *Lysosomes in Biology and Pathology*, *4* (North Holland/American Elsevier, New York), ch. 2

Bonhomme A., Fonty, G. and Senaud, J. (1982) 'Obtention de *Polyplastron multivesiculatum* (Cilié Entodiniomorphe du Rumen) en condition axénique', *Journal of Protozoology*, *29*, 231-3

Borror, A.C. (1968) 'Ecology of interstitial ciliates', *Transactions of American Microscopical Society*, *87*, 233-43

—(1980) 'Spatial distribution of marine ciliates: microecologic and biogeographic aspects of protozoan ecology', *Journal of Protozoology*, *27*, 10-13

Bovee, E.C. (1973) 'Preliminary report on the amoebae of the Atlantic Coast from Virginia to Massachusetts including the New York Bight', *Contract Report, U.S. Dept. Commer. Natl. Mar. Fish. Serv.* (Sandy Hook, New Jersey), 19pp

—(1979) 'Preliminary report on amoebas found in marine coastal waters of Virginia, New Jersey and Massachusetts', *Journal of Protozoology*, *26*, 26A

—and Sawyer, T.K. (1979) 'Marine flora and fauna of the north-eastern United States. Protozoa: Sarcodina: Amoebae', *U.S. Dept. Commer. Natl. Mar. Fish. Serv. Circ.*, 419

Bowers, B. (1977) 'Comparison of pinocytosis and phagocytosis in *Acanthamoeba castellani*', *Experimental Cell Research*, *110*, 409-17

—and Olszewski, T.E. (1972) 'Pinocytosis in *Acanthamoeba castellani*', *Journal of Cell Biology*, *53*, 681-94

Bowman, I.B.R. and Flynn, I.W. (1976) 'Oxidative metabolism of Trypanosomes', in W.H.R. Lumsden and D.A. Evans (eds.), *The Biology of the Kinetoplastida*, vol. 1, (Academic Press, London), ch. 10

Bradbury, P.C. (1966) 'The fine structure of the mature tomite of *Hyalophysa chattoni*', *Journal of Protozoology*, *13*, 591-607

—(1973) 'The fine structure of the cytostome of the apostomatous ciliate *Hyalophysa chattoni*', *Journal of Protozoology*, *20*, 405-14

—and Olive, L.S. (1980) 'Fine structure of the feeding stage of a sorogenic ciliate *Sorogena stoianovitchae* gen.n., sp.n.', *Journal of Protozoology*, *27*, 267-77

Brafield, A.E. (1964) 'The oxygen content of interstitial water in sandy shores', *Journal of Animal Ecology*, *33*, 97-116

Brand, T. von (1973) *Biochemistry of Parasites*, 2nd edn. (Academic Press, New York)

Brink, N. (1967) 'Ecological studies in biological filters', *Int.Revue ges. Hydrobiol. Hydrogr.*, *52*, 51-122

Brock, T.D. (1978) *Thermophilic Micro-organisms and Life at High Temperatures* (Springer-Verlag, New York)

Brooker, B.E. (1971) 'Fine structure of *Bodo saltans* and *Bodo caudatus* (Zoomastigophora: Protozoa) and their affinities with the Trypanosomatidae', *Bulletin of British Museum (Nat.Hist.)*, *22*, 82-102

Brown, M.G. (1940) 'Growth of protozoan cultures 2. *Leucophrys patula* and *Glaucoma pyriformis* in bacteria-free medium', *Physiological Zoology*, *13*, 277-82

Brown, T.J. (1964) 'The inter-relationship of two dominant ciliate species in an activated sludge plant with domestic effluent', *Journal of Protozoology*, *11* (Suppl.), 38

—(1965) 'A study of the protozoa in a diffused-air activated sludge plant', *Water Pollution Control*, *64*, 375-8

Browning, J.L. and Nelson, D.L. (1976) 'Biochemical studies of the excitable membrane of *Paramecium aurelia*', *Biochim. Biophys. Acta*, *448*, 338-51

Buetow, D.E. (1968) *The Biology of Euglena*, vols. 1, 2 (Academic Press, New York)

—(1982) *The Biology of Euglena*, vol. 3 (Academic Press, New York)

Bunt, J.S. (1963) 'Diatoms of Antarctic sea-ice as agents of primary production', *Nature (Lond.)*, *199*, 1255-7

—and Wood, E.J.F. (1963) 'Microalgae and Antarctic sea-ice', *Nature (Lond.)*, *199*, 1254-5

Burbanck, W.D. and Spoon, D.M. (1967) 'The use of sessile ciliates collected in plastic petri dishes for rapid assessment of water pollution', *Journal of Protozoology*, *14*, 739-44

Burkovsky, I.V. (1978) 'Structure, dynamics and production of a community of marine psammophilous ciliates', *Zool. Zhurnal*, *57*, 325-37

Burzell, L.A. (1973) 'Observations on the proboscis-cytopharynx and flagella in *Rhynchomonas metabolita* Pshenin 1964 (Zoomastigophora Bodonidae)', *Journal of Protozoology*, *20*, 385-93

—(1975) 'Fine structure of *Bodo curvifilus* Griessmann', *Journal of Protozoology*, *22*, 35-9

Butzel, H.M. and Bolten, A.B. (1968) 'The relationship of the nutritive state of the prey organism *Paramecium aurelia* to the growth and encystment of *Didinium nasutum*', *Journal of Protozoology*, *15*, 256-8

Byers, T.J., Akins, R.A., Maynard, B.J., Lefken, R.A. and Martin, S.M. (1980) 'Rapid growth of *Acanthamoeba* in defined medium: induction of encystment by glucose-acetate starvation', *Journal of Protozoology*, *27*, 216-19

Cairns, J.Jr. (1982) *Artificial Substrates* (Ann Arbor Science Publishers Inc., Michigan), ch. 2

—, Dahlberg, M.L., Dickson, K.L., Smith, N. and Waller, W.T. (1969) 'The relationship of freshwater protozoan communities to the MacArthur-Wilson equilibrium model', *American Naturalist*, *103*, 439-54

—, Plafkin, J.L., Yongue, W.H. Jr. and Kaesler, R.L. (1976a) 'Colonization of artificial substrates by protozoa: replicate samples', *Archiv für Protistenkunde*, *118*, 259-67

Cairns, J.Jr., Kaesler, R.L., Kuh, D.L., Plafkin, J.L. and Yongue, W.H. Jr. (1976b) 'The influence of natural perturbation upon protozoan communities inhabiting artificial substrates', *Transactions of American Microscopical Society*, *95*, 646-53

Campbell, A.S. (1942) 'The oceanic Tintinnoina of the plankton gathered during the last cruise of the Carnegie', *Publications of the Carnegie Institute, no. 537*, 163pp

Canning, E.U. (1975) *The Microsporidian Parasites of Platyhelminths – their Morphology, Development, Transmission and Pathogenicity* (C.I.H. Miscellaneous Publications, no. 2, Commonwealth Agricultural Bureau)

—, Foon, L.P. and Joe, L.K. (1974) 'Microsporidian parasites of trematode larvae from aquatic snails in West Malaysia', *Journal of Protozoology*, *21*, 19-25

Canter, H.M. and Lund, J.W.G. (1968) 'The importance of protozoa in controlling the abundance of planktonic algae in lakes', *Proceedings of the Linnean Society, London*, *179*, 203-19

Carpenter, K.E. (1928) *Life in Inland Waters* (Sidgwick and Jackson, London)

Cavalier-Smith, T. (1978) 'The evolutionary origin and phylogeny of microtubules, mitotic spindles and eukaryote flagella', *Biosystems*, *10*, 93-114

Chapman-Andresen, C. (1962) 'Studies on pinocytosis in amoebae', *Compt.Rend. Trav.Lab. Carlsberg*, *33*, 73-173, reprinted by Danish Science Press, Copenhagen

—(1973) 'Endocytotic processes' in K.W. Jeon (ed.), *The Biology of Amoeba* (Academic Press, New York), ch. 10

—and Holter, H. (1955) 'Studies on the ingestion of C^{14} glucose by pinocytosis in the amoeba *Chaos chaos*', *Experimental Cell Research*, *Suppl.3*, 52-63

—and Nilsson, J.R. (1967) 'Studies on endocytosis in amoebae: the distribution of pinocytotically ingested dyes in relation to food vacuoles in *Chaos chaos*', *Compt.Rend.Trav.Lab. Carlsberg*, *36*, 189-207

Chen, Y.T. (1950) 'The biology of *Peranema trichophorum*', *Quarterly Journal of Microscopical Science*, *91*, 279-308

Christiansen, R.G. and Marshall, J.M. (1965) '*Chaos chaos*, a study of phagocytosis', *Journal of Cell Biology*, *25*, 443-57

Clarke, R.T.J. (1977) 'Protozoa in the rumen ecosystem' in R.T.J. Clarke and T. Bauchop (eds.), *Microbial Ecology of the Gut* (Academic Press, London), pp. 251-75

Clay, E. (1964) 'The fauna and flora of sewage processes. 2. Species list', *I.C.I. Ltd Paints Division, Research Memorandum* PVM 45/a/732

Cleveland, L.R. (1925) 'The method by which *Trichonympha campanula*, a protozoan in the intestine of termites, ingests solid particles of wood for food', *Biological Bulletin*, *48*, 282-7

—and Grimstone, A.V. (1964) 'The fine structure of the flagellate *Mixotricha paradoxa* and its associated micro-organisms', *Proceedings of the Royal Society of London, Series B*, *157*, 668-83

—, Hall, S.R., Sanders, E.P. and Collier, J. (1934) 'The wood-eating roach, *Cryptocercus*, its protozoa, and the symbiosis between protozoa and roach', *Memorandum American Academy of Arts and Sciences*, *17*, i-x

Coleman, G.S. (1969) 'The metabolism of starch, maltose, glucose and some other sugars by the rumen ciliate, *Entodinium caudatum*', *Journal of General*

Microbiology, *57*, 303-32
Coleman, G.S. (1972) 'The metabolism of starch, glucose, amino acids, purines, pyrimidines, and bacteria by the rumen ciliate *Entodinium simplex*', *Journal of General Microbiology*, *71*, 117-31
—(1975) 'The inter-relationship between rumen ciliate protozoa and bacteria', in I.W. Macdonald and A.C.I. Warner (eds.), *Digestion and Metabolism in the Ruminant*, Proceedings 4th International Symposium on Ruminant Physiology (University of New England Publications, Australia)
— (1979) 'Rumen ciliate protozoa', in M. Levandowsky and S.H. Hutner (eds.), *Biochemistry and Physiology of Protozoa*, vol. 2 (Academic Press, New York), pp. 381-408
—, Davies, J.I. and Cash, M.A. (1972) 'The cultivation of the rumen ciliates *Epidinium ecaudatum caudatum* and *Polyplastron multivesiculatum in vitro*', *Journal of General Microbiology*, *73*, 509-21
—and Laurie, J.I. (1974) 'The metabolism of starch, glucose, amino acids, purines, pyrimidines, and bacteria by three *Epidinium* spp. isolated from the rumen', *Journal of General Microbiology*, *85*, 244-56
—and — (1976) 'The uptake and metabolism of glucose, maltose and starch by the rumen ciliate *Epidinium ecaudatum caudatum*', *Journal of General Microbiology*, *95*, 364-74
—, —, Bailey, J.E. and Holdgate, S.A. (1976) 'The cultivation of cellulolytic protozoa isolated from the rumen', *Journal of General Microbiology*, *95*, 144-57
Coles, A.M., Swoboda, B.E.P. and Ryley, J.F. (1980) 'Thymidylate synthetase as a chemotherapeutic target in the treatment of avian coccidiosis', *Journal of Protozoology*, *27*, 502-6
Conner, R.L., Landrey, J.R. and Czarkowski, N. (1982) 'The effect of specific sterols on cell size and fatty acid composition of *Tetrahymena pyriformis* W', *Journal of Protozoology*, *29*, 105-9
Cook, J.R. (1968) 'The cultivation and growth of *Euglena*' in D.E. Buetow (ed.), *The Biology of Euglena*, vol. 1 (Academic Press, New York)
Corliss, J.O. (1973) 'Evolutionary trends in patterns of stomatogenesis in the ciliate protozoa', *Journal of Protozoology*, *20*, 506
—(1974) 'Remarks on the composition of the large ciliate class Kinetofragminophora, de Puytorac *et al.* 1974', *Journal of Protozoology*, *21*, 207-20
— (1979) *The Ciliated Protozoa*, 2nd edn. (Pergamon Press, New York)
Cosgrove, W.B. and Swanson, B.K. (1952) 'Growth of *Chilomonas paramecium* in simple organic media', *Physiological Zoology*, *25*, 287-92
Curds, C.R. (1963) 'The Flocculation of suspended matter by *Paramecium caudatum*', *Journal of General Microbiology*, *33*, 357-63
—(1966) 'An ecological study of the ciliated protozoa in activated sludge', *Oikos*, *15*, 282-9
—(1969) *An Illustrated Key to the British Freshwater Ciliated Protozoa Commonly Found in Activated Sludge* (HMSO, London)
— (1975) 'Protozoa' in C.R. Curds and H.A. Hawkes (eds.), *Ecological Aspects of Used-Water Treatment* 1. *The Organisms and their Ecology* (Academic Press, London), ch. 5
—and Cockburn, A. (1970a) 'Protozoa in biological sewage treatment processes. I. Survey of the protozoan fauna of British percolating filters and acti-

vated sludge plants', *Water Research*, *4*, 225-36

Curds, C.R. and — (1970b) 'Protozoa in biological sewage treatment processes. II.Protozoa as indicators in the activated sludge process', *Water Research*, *4*, 237-49

—and — (1971) 'Continuous monoxenic culture of *Tetrahymena pyriformis*', *Journal of General Microbiology*, *66*, 95-108

—, — and Vandyke, J.M. (1968) 'An experimental study of the role of ciliated protozoa in the activated sludge process', *Water Pollution Control*, *67*, 312-29

—and Vandyke, J.M. (1966) 'Feeding habits and growth rates of some freshwater ciliates found in activated sludge plants', *Journal of Applied Ecology*, *3*, 127-37

Current, W.L. (1979) '*Henneguya adiposa* Minchen (Myxosporida) in the Channel catfish: ultrastucture of the plasmodium wall and sporogenesis', *Journal of Protozoology*, *26*, 209-17

—and Janovy, J. (1976) 'Ultrastructure of interlamellar *Henneguya exilis* in the Channel catfish', *Journal of Parasitology*, *62*, 975-81

—and — (1978) 'Comparative study of the ultrastructure of interlamellar and intralamellar types of *Henneguya exilis* Kudo from Channel catfish', *Journal of Protozoology*, *25*, 56-65

—, — and Knight, S.A. (1979) '*Myxosoma funduli* Kudo in *Fundulus Kansae*: ultrastructure of the plasmodium wall and of sporogenesis', *Journal of Protozoology*, *26*, 574-83

Curry, A. and Butler, R.D. (1979) 'Giant forms of the suctorian ciliate *Discophrya collini*', *Protoplasma*, *100*, 125-37

Daley, R.J., Morris, G.P. and Brown, S.R. (1973) 'Phagotrophic ingestion of a blue-green alga by *Ochromonas*', *Journal of Protozoology*, *20*, 58-61

Danforth, H.D., Aikawa, M., Cochrane, A.H. and Nussenzweig, R.S. (1980) 'Sporozoites of mammalian malaria: attachment to, interiorization and fate of macrophages', *Journal of Protozoology*, *27*, 193-202

Danforth, W.F. (1968) 'Respiration' in D.E. Buetow (ed.), *The Biology of Euglena*, vol. 2 (Academic Press, New York), pp. 55-71

Davies, A.J. (1982) 'Further studies on *Haemogregarina bigemina* Laveran & Mesnil, the marine fish *Blennius pholis* L., and the isopod *Gnathia maxilliaris* Montagu', *Journal of Protozoology*, *29*, 576-83

Davis, P.G., Caron, D.M. and Sieburth, J.McN. (1978) 'Oceanic amoebae from the North Atlantic: culture, distribution and taxonomy', *Journal of American Microscopical Society*, *97*, 73-88

de Duve, C. (1963) *Lysosomes* (Ciba Foundation Symposium), (Churchill, London), p. 1

Deegan, T. and Maegraith, B.G. (1956) 'Studies on the nature of the malarial pigment (haemozoin) II', *Annals of Tropical Medicine and Parasitology*, *50*, 212-22

Dehority, B.A. (1979) 'Ciliate protozoa in the rumen of Brazilian water buffalo, *Bubalis bubalis* Linn.', *Journal of Protozoology*, *26*, 536-44

de Isola, E.L.D., Lammel, E.M., Katzin, V.J. and Gonzalez Cappa, S.M. (1981) 'Influence of organ extracts of *Triatoma infestans* on differentiation of *Trypanosoma cruzi*', *Journal of Parasitology*, *67*, 53-8

Dewey, V.C. and Kidder, G.W. (1940) 'Growth studies on ciliates VI diagnosis, sterilization and growth characteristics of *Perspira ovum*', *Biological Bulletin*, *79*, 255-71

Dive, D. (1975) 'Influence de la concentration bactérienne sur la croissance de *Colpidium campylum*', *Journal of Protozoology*, *22* 545-50

Dobell, C. (1960) *Antony van Leeuwenhoek and his Little Animals* (Dover Publications Inc., New York)

Dodge, J.D. and Crawford, R.M. (1971) 'A fine structural survey of dinoflagellate pyrenoids and food reserves', *Botanical Journal of Linnean Society*, *64*, 105-15

Dogiel, V.A. (1965) *General Protozoology*, 2nd edn. (Oxford University Press, London)

Dragesco, J. (1960) 'Les ciliés mésopsammiques littoraux', *Tra. Sta. Biol. Roscoff*, 1-356

Droop, M.R. (1954) 'A note on the isolation of small marine algae and flagellates in pure cultures', *Journal of Marine Biological Association, UK*, *33*, 511-14

—(1959) 'Water-soluble factors in the nutrition of *Oxyrrhis marina*', *Journal of Marine Biological Association, UK*, *38*, 605-20

Dubowsky, N. (1974) 'Selectivity of ingestion and digestion in the chrysomonad flagellate *Ochromonas malhamensis*', *Journal of Protozoology*, *21*, 295-8

Dunham, P.B. and Kropp, D.L. (1973), 'Regulation of salts and water in *Tetrahymena*' in A.M. Elliott (ed.), *Biology of Tetrahymena* (Dowden, Hutchinson and Ross, Inc., Pennsylvania), ch. 6

Dunlap, M. (1977) 'Localization of calcium channels in *Paramecium caudatum*', *Journal of Physiology*, *271*, 119-33

Duysens, L.N.M. and Amesz, J. (1962) 'Function and identification of two photochemical systems in photosynthesis', *Biochim. Biophys. Acta.*, *64*, 243-60

Eadie, J.M. (1967) 'Studies on the ecology of certain rumen ciliate protozoa', *Journal of General Microbiology*, *49*, 175-94

Elliott, A.M. and Clemmons, G.L. (1966) 'An ultrastructural study of ingestion and digestion in *Tetrahymena pyriformis*', *Journal of Protozoology*, *13*, 311-23

Evans, W.R. (1968) 'Photosynthesis in *Euglena*' in D.E. Buetow (ed.), *The Biology of Euglena*, vol. 2 (Academic Press, New York), pp. 73-84

Eyden, B.P. and Vickerman, K. (1975) 'Ultrastructure and vacuolar movements in the free-living diplomonad *Trepomonas agilis*', *Journal of Protozoology*, *22*, 54-66

Fagundes, L.J.M., Angluster, J., Gibert, B. and Roitman, I. (1980) 'Synthesis of sterols in *Herpetomonas samuelpessoai*: influence of growth conditions', *Journal of Protozoology*, *27*, 238-41

Farley, C.A. (1975) 'Epizootic and enzootic aspects of *Minchina nelsoni* (Haplosporida) disease in Maryland oysters', *Journal of Protozoology*, *22*, 418-27

Farmer, J.N. (1980) *The Protozoa* (C.V. Mosby Company, St Louis)

Fauré-Fremiet, E. (1924) 'Contribution à la connaissance des infusoires planktoniques', *Bull.Biol. France Belgique (Suppl.)*, *6*, 1-171

—(1950) 'Écologie des ciliés psammophiles littoraux', *Bull.Biol.France Belgique*, *84*, 35-75

—(1951a) 'Associations infusoriennes à *Beggiatoa*', *Hydrobiologie*, *3*, 65-71

—(1951b) 'The marine and sand-dwelling ciliates of Cape Cod', *Biological Bulletin*, *100*, 59-70

Fenchel, T. (1968) 'The ecology of marine microbenthos. II. The food of marine benthic ciliates', *Ophelia*, *5*, 73-121

Fenchel, T. (1969) 'The ecology of marine microbenthos. IV. Structure and function of the benthic ecosystem', *Ophelia*, *6*, 1-182

—(1972) 'Aspects of decomposer food chains in marine benthos', *Verh.Deutsch. Zool. Ges. 65 Jahresversamml.*, *14*, 14-22

—(1980a) 'Relation between particle size selection and clearance in suspension-feeding ciliates', *Limnology and Oceanography*, *25*, 733-8

—(1980b) 'Suspension feeding in ciliated protozoa: functional response and particle size selection', *Microbial Ecology*, *6*, 1-11

—(1980c) 'Suspension feeding in ciliated protozoa: feeding rates and their ecological significance', *Microbial Ecology*, *6*, 13-25

—(1980d) 'Suspension feeding in ciliated protozoa: structure and function of feeding organelles', *Archiv für Protistenkunde*, *123*, 239-60

—and Harrison, P. (1976) 'The significance of bacterial grazing and mineral cycling for the decomposition of particulate detritus', in J.M. Anderson and A. Macfadyen (eds.), *The Role of Terrestrial and Aquatic Organisms in Decomposition Processes* (Blackwell, London), pp. 285-99

—and Lee, C.C. (1972) 'Studies on ciliates associated with sea ice from Antarctica. I. The nature of the fauna', *Archiv für Protistenkunde*, *114*, 231-6

—, Perry, T. and Thane, A. (1977) 'Anaerobiosis and symbiosis with bacteria in free-living ciliates', *Journal of Protozoology*, *24*, 154-63

—and Small, E.B. (1980) 'Structure and function of the oral cavity and its organelles in the hymenostome ciliate *Glaucoma*', *Transactions of American Microscopical Society*, *99*, 52-60

Finlay, B.J. (1978) 'Community production and respiration by ciliated protozoa in the benthos of a small eutrophic loch', *Freshwater Biology*, *8*, 327-41

—(1980) 'Temporal and vertical distribution of ciliophoran communities in the benthos of a small eutrophic loch with particular reference to the redox profile', *Freshwater Biology*, *10*, 15-34

—, Bannister, P. and Stewart, J. (1979) 'Temporal variation in benthic ciliates and the application of association analysis', *Freshwater Biology*, *9*, 45-53

Finley, H.E. (1930) 'Toleration of freshwater protozoa to increased salinity', *Ecology*, *11*, 337-46

Fischer-Defoy, D. and Hausmann, K. (1977) 'Untersuchungen zur phagocytose bei *Climacostomum virens*', *Protistologica*, *13*, 459-76

Fisher, S.G. and Likens, G.E. (1973) 'Energy flow in Bear Brook, New Hampshire: an integrative approach to stream ecosystem metabolism', *Ecological Monographs*, *43*, 421-39

Flint, E.A. and Stout, J.D. (1960) 'Microbiology of some soils from Antarctica', *Nature (London)*, *188*, 767-8

Flynn, I.W. and Bowman, I.B.R. (1973) 'The metabolism of carbohydrate by pleomorphic African trypanosomes', *Comp. Biochem. Physiol.*, *45B*, 24-42

Fok, A.K., Allen, R.D. and Kaneshiro, E.S. (1981) 'Axenic *Paramecium caudatum*. III.Biochemical and physiological changes with culture age', *European Journal of Cell Biology*, *25*, 193-201

—, Lee, Y. and Allen, R.D. (1982) 'The correlation of digestive vacuole pH and size with the digestive cycle in *Paramecium caudatum*', *Journal of Protozoology*, *29*, 409-14

Forsee, W.J. and Kahn, J.S. (1972) 'Carbon dioxide fixation by isolated chloroplasts of *Euglena gracilis*', *Archiv. Biochemistry & Biophysics*, *150*, 302-9

French, C.S. (1971) 'The distribution and action in photosynthesis of several forms of chlorophyll', *Proc. Natn. Acad. Sci. USA*, *68*, 2893-7

Fulton, J.D. (1969) 'Metabolism and pathogenic mechanism in parasitic protozoa' in T.T. Chen (ed.), *Research in Protozoology*, vol. 3 (Pergamon Press, New York), pp. 389-504

—and Spooner, D.F. (1957) 'Preliminary observations on the metabolism of *Toxoplasma gondii*', *Trans. Roy.Soc.Trop.Med.Hyg.* *51*, 123-5

—and — (1960) 'Metabolic studies on *Toxoplasma gondii*', *Experimental Parasitology*, *9*, 293-301

Gasith, A. and Hasler, A.D. (1976) 'Airborne litterfall as a source of organic matter in lakes', *Limnology and Oceanography*, *21*, 253-8

Gause, G.F. (1934) *The Struggle for Existence* (Williams & Wilkins, Baltimore)

—, Nastukova, O.K. and Alpatov, W.W. (1934) 'The influence of biologically conditioned media on the growth of a mixed population of *Paramecium caudatum* and *P. aurelia*', *Journal of Animal Ecology*, *3*, 222-30

—, Smaragdova, N.P. and Witt, A.A. (1936) 'Further studies of interaction between predator and prey', *Journal of Animal Ecology*, *5*, 1-18

Gelei, J. von (1934) 'Der feinere Bau des Cytopharynx von *Paramecium* und seine systematische Bedeutung', *Archiv für Protistenkunde*, *82*, 331-62

Gerlach, S.A. (1971) 'On the importance of marine meiofauna for benthos communities', *Oecologia*, *6*, 176-90

Gibbs, S.P. (1960) 'The fine structure of *Euglena gracilis* with special reference to chloroplasts and pyrenoids', *Journal of Ultrastructure Research*, *4*, 127-48

—(1962a) 'The ultrastructure of the pyrenoids of green algae', *Journal of Ultrastructure Research*, *7*, 262-72

—(1962b) 'The ultrastructure of chloroplasts of algae', *Journal of Ultrastructure Research*, *7*, 418-35

Giese, A.C. (1973) Blepharisma: *the Biology of a Light-sensitive Protozoan* (Stanford University Press, Stanford, California)

Gomori, G. (1952) *Microscopic Histochemistry: Principles and Practice* (University of Chicago Press, Chicago), pp. 189-94

Gorham, E. (1958) 'Observations on the formation and breakdown of the oxidized microzone at the mud surface in lakes', *Limnology and Oceanography*, *3*, 291-8

Goulder, R. (1971a) 'Vertical distribution of some ciliated protozoa in two freshwater sediments', *Oikos*, *22*, 199-203

—(1971b) 'The effects of saprobic conditions on some ciliated protozoa in the benthos and hypolimnion of a eutrophic pond', *Freshwater Biology*, *1*, 307-18

—(1972) 'Grazing by the ciliated protozoan *Loxodes magnus* on the alga *Scenedesmus* in a eutrophic pond', *Oikos*, *23*, 109-15

—(1974) 'The seasonal and spatial distribution of some benthic ciliated protozoa of Esthwaite Water', *Freshwater Biology*, *4*, 127-47

Govindjee, R. and Braun, B.Z. (1974) 'Light absorption, emission and photosynthesis' in W.D.P. Stewart (ed.), *Algal Physiology and Biochemistry* (Blackwell, Oxford), ch. 12

Graham, H.W. (1942) 'Studies in the morphology, taxonomy and ecology of Peridiniales', *Carnegie Institute Publication*, *no. 542*, pp. 129

Grell, K.G. (1973) *Protozoology* (Springer-Verlag, Berlin)

Grim, J.N. (1967) 'Ultrastructure of pellicular and ciliary structures in *Euplotes eurystomus*', *Journal of Protozoology*, *14*, 625-34

Grimes, B.H. (1976) 'Notes on the distribution of *Hyalophysa* and *Gymnodinioides* on crustacean hosts in coastal North Carolina and a description of *Hyalophysa trageri* sp.n.', *Journal of Protozoology*, *23*, 246-51

Gross, J.A. and Jahn, T.L. (1962) 'Cellular responses to thermal and photostress. I. *Euglena and Chlamydomonas*', *Journal of Protozoology, 9*, 340-6

Grula, M. and Bovee, E.C. (1977) 'Ingestion and subsequent loss of a rotifer by *Stentor coeruleus*', *Transactions of American Microscopical Society*, *96*, 538-9

Güde, H. (1979) 'Grazing by protozoa as selection factor for activated sludge bacteria', *Microbial Ecology*, *5*, 225-37

Gutierrez, J. (1955) 'Experiments on the culture and physiology of holotrichs from the bovine rumen', *Biochemical Journal*, *60*, 516-22

Gutteridge, W.E. and Coombs, G.H. (1977) *Biochemistry of Parasitic Protozoa* (Macmillan Press, Ltd, London)

— and Rogerson, G.W. (1979) 'Biochemical aspects of the biology of *Trypanosoma cruzi*', in W.H.R. Lumsden and D.A. Evans (eds.), *Biology of the Kinetoplastida*, vol. 2 (Academic Press, London), pp. 619-52

Hamilton, R.D. and Preslan, J.E. (1970) 'Observations on the continuous culture of a planktonic phagotrophic protozoan', *Journal of Experimental Marine Biology and Ecology*, *5*, 94-104

Hammond, D.M. and Long, P.L. (1973) *The Coccidia*: Eimeria, Isospora *and* Toxoplasma *and Related Genera* (University Park Press, Baltimore)

—, Scholtyseck, E. and Chobotar, B. (1967) 'Fine structures associated with nutrition of the intracellular parasite *Eimeria auburnensis*', *Journal of Protozoology*, *14*, 678-83

—, — and — (1969) 'Fine structural study of the microgametogenesis of *Eimeria auburnensis*', *Z. Parasitenk.*, *33*, 65-84

Hanel, K. (1979) 'Systematik und Okologie der farblosen Flagellaten des Abwassers', *Archiv für Protistenkunde*, *121*, 73-137

Hanson, E.D. (1977) *The Origin and Early Evolution of Animals* (Wesleyan University Press, Connecticut)

Hara, R. and Asai, H. (1980) 'Electrophysiological responses of *Didinium nasutum* to *Paramecium* capture and mechanical stimulation', *Nature*, *283*, 869-70

Harrison, P.G. and Mann, K.H. (1975) 'Detritus formation from eelgrass (*Zostera marina* L.): the relative effects of fragmentation, leaching and decay', *Limnology and Oceanography*, *20*, 924-34

Hart, D.T. and Coombs, G.H. (1982) 'The energy-producing pathways of *Leishmania mexicana*', *Journal of Protozoology*, *29*, 320

Hartwig, E. (1973) 'Die ciliaten des Gezeiten-Sand strandes de Nordsee insel Sylt. II. Okologie', *Akad. Wiss. Lit. Math-Nat. Kl. Mikrofauna Meeresbodens*, *21*, 1-171

—(1977) 'On the interstitial ciliate fauna of Bermuda', *Cah. Biol. Mar. 18*, 113-26

—and Parker, J.G. (1977) 'On the systematics and ecology of interstitial ciliates of sandy beaches in North Yorkshire', *Journal of Marine Biological Association, UK*, *57*, 735-60

Hauser, M. and van Eys, H. (1976) '*Heliophrya erhardi*, microtubules and microfilaments in suctorian tentacles', *Journal of Cell Science*, *20*, 589-93

Hausman, L.A. (1923) 'Studies on the fauna of sprinkling filter beds and Imhoff tanks', *American Journal of Public Health*, *3*, 656-85

Hausmann, K. and Patterson, D.J. (1982) 'Feeding in *Actinophrys*. II. Pseudopod formation and membrane production during prey capture by a heliozoan', *Cell Motility*, *2*, 9-24

—and Peck, R. (1978) 'Microtubules and microfilaments as major components of a phagocytic apparatus: the cytopharyngeal basket of the ciliate *Pseudomicrothorax dubius*', *Differentiation*, *11*. 157-67

—and — (1979) 'The mode of function of the pharyngeal basket of the ciliate *Pseudomicrothorax dubius*', *Differentiation*, *14*, 147-58

Hawkes, H.A. (1963) *The Ecology of Waste Water Treatment* (Pergamon Press, Oxford)

Heal, O.W. (1962) 'The abundance and microdistribution of testate amoebae (Rhizopoda: Testacea) in Sphagnum', *Oikos*, *13*, 35-47

—(1965) 'Observations on testate amoebae (Protozoa: Rhizopoda) from Signy Island, South Orkney Islands', *British Antarctic Survey Bulletin*, *no. 6*, 43-7

—, Bailey, A.D. and Latter, P.M. (1967) 'Bacteria, fungi and protozoa in Signy Island soils compared with those from a temperate moorland', *Phil. Trans. Roy. Soc., Series B.*, *252*, 191-7

Henebry, M.S. and Cairns, J.Jr. (1980a), 'The effect of source pool maturity on the process of island colonization: an experimental approach with protozoan communities', *Oikos*, *35*, 107-14

—and — (1980b) 'Monitoring of stream pollution using protozoan communities on artificial substrates', *Transactions of American Microscopical Society*, *99*, 151-60

Hewett, S.W. (1980) 'Prey-dependent cell size in a protozoan predator', *Journal of Protozoology*, *27*, 311-13

Hibberd, D.J. (1975) 'Observations on the ultrastructure of the choanoflagellate *Codosiga botrytis* (Ehr.) Saville-Kent with special reference to the flagellar apparatus', *Journal of Cell Science*, *17*, 191-219

Hill, D.L. (1972) *The Biochemistry and Physiology of* Tetrahymena (Academic Press, New York), ch. 1

Hill, R. and Bendall, F. (1960) 'Function of two cytochrome components in chloroplasts: a working hypothesis', *Nature (Lond.)*, *186*, 136-7

Hindle, E. (1932) 'Some new thermophilic organisms', *Journal of the Royal Microscopical Society*, *52*, 123-33, Trans.VII

Hitchin, E.T. and Butler, R.D. (1973) 'Ultrastructural studies of the commensal suctorian *Choanophrya infundibulifera* Hartog.', *Zeit. Zellforsch.*, *144*, 37-57

—and — (1974) 'The ultrastructure and function of the tentacle in *Rhyncheta cyclopum* Zenker', *Journal of Ultrastructure Research*, *46*, 279-95

Hoare, C.A. (1964) 'Morphological and taxonomic studies on mammalian trypanosomes. X. Revision of the systematics', *Journal of Protozoology*, *11*, 206-7

—and Wallace, F.G. (1966) 'Developmental stages of trypanosomatid flagellates: new terminology', *Nature (Lond.)*, *212*, 1385-6

Hogg, J.F. and Kornberg, H.L. (1963) 'The metabolism of C_2-compounds in micro-organisms. 9. The glyoxylate cycle in *Tetrahymena*', *Biochemical Journal*, *86*, 462-8

Holz, G.G. Jr. (1954) 'The oxidative metabolism of a cryptomonad flagellate *Chilomonas paramecium*', *Journal of Protozoology*, *1*, 114-20
Honigberg, B.M. (1967) 'Chemistry of parasitism among some protozoa' in G.W. Kidder (ed.), *Chemical Zoology, I. Protozoa* (Academic Press, New York), pp. 695-801
Howard, B.H. (1959) 'The biochemistry of rumen protozoa. 1. Carbohydrate fermentation by *Dasytricha* and *Isotricha*', *Biochemical Journal*, *71*, 671-5
Hull, R.S. (1961a) 'Studies on suctorian protozoa: the mechanism of prey adherence', *Journal of Protozoology*, *8*, 343-50
—(1961b) 'Studies on suctorian protozoa: the mechanism of ingestion of prey cytoplasm', *Journal of Protozoology*, *8*, 351-9
Hungate, R.E. (1939) 'Experiments on the nutrition of *Zootermopsis*', *Ecology*, *20*, 230-5
—(1942) 'The culture of *Entodinium neglectum*, with experiments on the digestion of cellulose', *Biological Bulletin*, *83*, 303-19
—(1943a) 'Further experiments on the cellulose digestion by protozoa in the rumen of cattle', *Biological Bulletin*, *84*, 157-63
—(1943b) 'Quantitative analyses of the cellulose fermentation by termite protozoa', *Annals Entomological Society of America*, *36*, 730-9
— (1955) 'Mutualistic intestinal protozoa' in S.H. Hutner and A. Lwoff (eds.), *Biochemistry and Physiology of Protozoa*, vol. II (Academic Press, New York), pp. 159-99
—(1966) *The Rumen and its Microbes* (Academic Press, New York)
—(1978) 'The rumen protozoa' in J.P. Kreier (ed.), *Parasitic Protozoa* (Academic Press, New York), pp. 684-7
Hunter, F.R. and Lee, J.W. (1962) 'On the metabolism of *Astasia longa*', *Journal of Protozoology*, *9*, 74-8
Hutner, S.H., Bacchi, C.J. and Baker, H. (1979) 'Nutrition of the Kinetoplastida' in W.H.R. Lumsden and D.A. Evans (eds), *Biology of the Kinetoplastida*, vol. 2 (Academic Press, London), pp. 653-92
—and Provasoli, L. (1955) 'Comparative biochemistry of flagellates' in S.H. Hutner and A. Lwoff (eds.), *Biochemistry and Physiology of Protozoa*, vol. 2 (Academic Press, New York), pp. 17-43
Issel, R. (1906) 'Sulla termibiose negli animali acquatici', *Atti.Soc.Ligust.Sci.Nat. Geog.*, *17*, 3-72
—(1910) 'La faune des sources thermales de Viterbo', *Int.Rev.des ges. Hydrobiologie*, *3*, 178-80
Janasch, H.W. and Wirsen, C.O. (1973) 'Deep-sea micro-organisms: *in situ* response to nutrient enrichment', *Science*, *180*, 641-3
Jeon, K.W. (1973) *The Biology of Amoeba* (Academic Press, New York)
Johannes, R.E. (1965) 'Influence of marine protozoa on nutrient regeneration', *Limnology and Oceanography*, *10*, 434-42
Johnson, J.G., Epstein, N., Shiroishi, T. and Miller L.H. (1981) 'Identification of surface proteins on viable *Plasmodium knowlesi* merozoites', *Journal of Protozoology*, *28*, 160-4
Jones, T.C. (1980) 'Immunopathology of Toxoplasmosis', *Immunopathology*, *2*, 387-97
—and Hirsch, J.G. (1972) 'The interaction between *Toxoplasma gondii* and mammalian cells II', *Journal of Experimental Medicine*, *136*, 1173-94

Josephson, S.L., Weik, R.R. and John, D.T. (1977) 'Concanavalin A-induced agglutination in *Naegleria*', *Amer.J.Trop.Med.Hyg.*, *26*, 856-8

Jurand, A. (1961) 'An electron microscope study of food vacuoles in *Paramecium aurelia*', *Journal of Protozoology*, *8*, 125-30

Kahan, D. (1972) '*Cyclidium citrullus* Cohn, a ciliate from the hot springs of Tiberias (Israel)', *Journal of Protozoology*, *19*, 593-7

Kahl, A. (1930-5) 'Urtiere oder Protozoa. I.Wimpertiere oder Ciliata (Infusoria)' in F. Dahl (ed.), *Die Tierwelt Deutschlands*, Teil 18, 21, 25 an 30 (G. Fischer, Jena)

Kaneda, M. (1960) 'Phase contrast microscopy of cytoplasmic organelles in the gymnostome ciliate *Chlamydodon pedarius*', *Journal of Protozoology*, *7*, 306-13

Kaneshiro, E.S., Beischel, L.S., Merkel, S.J. and Rhoads, D.E. (1979) 'The fatty acid composition of *Paramecium aurelia* cells and cilia: changes with culture age', *Journal of Protozoology*, *26*, 147-58

Karakashian, M.W. (1975) 'Symbiosis in *Paramecium bursaria*', in D.H. Jennings and D.L. Lee (eds.), *Symbiosis*, Society of Experimental Biology, Symp. 29 (Cambridge University Press)

Karakashian, S.J. and Rudzinska, M.A. (1981) 'Inhibition of lysosomal fusion with symbiont-containing vacuoles in *Paramecium bursaria*', *Experimental Cell Research*, *131*, 387-93

Kareem, H.A. and Soldo, A.T. (1978) 'Glycogen in the marine protozoan *Parauronema acutum*', *Journal of Protozoology*, *25*, 560-2

Karpenko, A.A., Railkin, A.I. and Seravin, L.N. (1977) 'Feeding behaviour of unicellular animals. 2. The role of prey mobility in the feeding behaviour of protozoa', *Acta Protozoologica*, *16*, 333-4

Kawabata, A., Miyatake, K. and Kitaoka, S. (1982) 'Effect of temperature on the contents of the two energy-reserve substances, paramylon and wax esters, in *Euglena gracilis*', *Journal of Protozoology*, *29*, 421-3

Khavkin, T. (1981) 'Histological and ultrastructural studies of the interaction of *Toxoplasma gondii* tachyzoites with mouse omentum in experimental infection', *Journal of Protozoology*, *28*, 317-25

Khlebovich, T.V. (1976) 'Dependence on generation time and nutrition rate in the infusorian *Dileptus anser* on food concentration', *Tsitologiya Leningrad*, *18*, 109-12 (English Summary)

Kidder, G.W. and Dewey, V.C. (1951) 'The biochemistry of ciliates in pure culture' in A. Lwoff (ed.), *Biochemistry and Physiology of Protozoa* (Academic Press, New York), pp. 324-97

Kimball, R.F. and Prescott, D.M. (1962) 'Deoxyribonucleic acid synthesis and distribution during growth and amitosis of the macronucleus of *Euplotes*', *Journal of Protozoology*, *9*, 88-92

Klaveness, D. (1981) '*Rhodomonas lacustris*: ultrastructure of the vegetative cell', *Journal of Protozoology*, *28*, 83-90

Kofoid, C.A. and Skogsberg, T. (1928) 'The dinoflagellate: the dinophysidae', *Mem.Mus.Comp.Zool.Harvard*, *51*, 1-766

—and Swezy, O. (1921) 'The free-living unarmoured Dinoflagellata', *Memoirs Univ.Calif.Berkeley*, *5*, pp. 563

Kolkwitz, R. and Marsson, M. (1908) 'Okologie der pflanzlichen saprobien', *Ber.Deutsch bot. Ges. (A)*, *26*, 505-19

Koroly, M.J. and Conner, R.L. (1976) 'Unsaturated fatty acid biosynthesis in *Tetrahymena*', *Journal of Biological Chemistry*, *251*, 7588-92

Kreier, J.P. (ed.) (1980) *Malaria* (Academic Press, New York, 3 vols.

Kudo, R.R. (1966) *Protozoology*, 5th edn. (C.C. Thomas, Springfield, Ill.)

Kusamran, K., Mattox, S.M. and Thompson, G.A. (1980) 'Studies on the size, location and turnover of calcium pools accessible to growing *Tetrahymena* cells', *Biochim.Biophys.Acta*, *598*, 16-26

Lackey, J.B. (1925) 'The fauna of Imhoff tanks', *Bulletin N.J.Agr.Exp.Sta.*, *no. 417*

—(1932) 'Oxygen deficiency and sewage protozoa: with descriptions of some new species', *Biological Bulletin*, *63*, 287-95

—(1938) 'A study of some ecologic factors affecting the distribution of protozoa', *Ecological Monographs*, *8*, 501-27

—(1967) 'The microbiota of estuaries and their roles' in G.H. Lauff (ed.), *Estuaries* (AAAS Publications, 83), pp. 291-302

Lanar, D.E. (1979) 'Growth and differentiation of *Trypanosoma cruzi* cultivated with a *Triatoma infestans* embryo cell line', *Journal of Protozoology*, *26*, 457-62

Langreth, S.G. (1976) 'Extracellular feeding mechanisms in *Babesia microti* and *Plasmodium lophurae*', *Journal of Protozoology*, *23*, 215-23

Latzko, E. and Gibbs, M. (1969) 'Enzyme activities of the carbon reduction cycle in some photosynthetic organisms', *Plant Physiol. Lancaster*, *44*, 295-300

Laval, M. (1971) 'Ultrastructure et mode de nutrition du choanoflagellé *Salpinoeca pelgica* (sp.n.)', *Protistologica*, *7*, 325-36

Laybourn, J.E.M. and Stewart, J.M. (1975) 'Studies on consumption and growth in the ciliate *Colpidium campylum* Stokes', *Journal of Animal Ecology*, *44*, 165-74

Leadbeater, B.S.C. (1972) 'Fine structural observations on some marine choanoflagellates from the coast of Norway', *Journal of Marine Biological Association UK*, *52*, 67-79

—(1973) 'External morphology of some marine choanoflagellates from the coast of Yugoslavia', *Archiv. für Protistenkunde*, *115*, 234-52

—and Manton, I. (1974) 'Preliminary observations on the chemistry and biology of the lorica of a collared flagellate (*Stephanoeca diplocostata* Ellis)', *Journal of Marine Biological Association, UK*, *54*, 269-76

—and Morton, C. (1974) 'A microscopical study of a marine species of *Codosiga* James-Clark (Choanoflagellata)', *Biological Journal Linnean Society*, *6*, 337-47

Lebour, M.V. (1925) *The Dinoflagellates of the Northern Seas* (Marine Biological Association, UK, Plymouth), pp. 250

Lee, J.J. (1974) 'Towards understanding the niche of Foraminifera' in R.H. Hedley and C.G. Adams (eds.), *Foraminifera*, vol. 1 (Academic Press, London), pp. 207-60

—(1980a) 'Informational energy flow as an aspect of protozoan nutrition', *Journal of Protozoology*, *27*, 5-9

—(1980b) 'Nutrition and Physiology of the Foraminifera' in M. Levandowsky and S.H. Hutner (eds.), *Nutrition and Physiology of Protozoa*, vol. 3 (Academic Press, New York), pp. 43-66

Lee, J.J., Freudenthal, H.D., Kossay, V. and Bé,A. (1965) 'Cytological observations on two planktonic Foraminifera *Globigerina bulloides* and *Globigerinoides ruber*', *Journal of Protozoology*, *12*, 531-42
—, McEnery, M., Pierce, S., Freudenthal, H.D. and Muller, W.A. (1966a) 'Tracer experiments in feeding littoral foraminifera', *Journal of Protozoology*, *13*, 659-70
—, — , — and Muller, W.A. (1966b) 'Prey and predator relationships in the nutrition of certain littoral foraminifera', *Journal of Protozoology*, *13* (Suppl.), 23
Leedale, G.F. (1967) *Euglenoid Flagellates* (Prentice Hall, Inc., Englewood Cliffs, New Jersey)
—(1968) 'The nucleus in *Euglena*' in D.E. Buetow (ed.), *The Biology of Euglena*, vol. 1 (Academic Press, New York), pp. 185-242
Leeuwenhoek, A. van, see Dobell (1960)
Levine, N.D. (1972) 'Relationship between certain protozoa and other animals' in T-T.Chen (ed.), *Research in Protozoology*, vol. 4 (Pergamon Press, Oxford)
—(1978) '*Perkinsus* gen.n. and other new taxa in the protozoan phylum Apicomplexa', *Journal of Parasitology*, *64*, 549
—(1982) 'Some corrections in Haemogregarine (Apicomplexa: Protozoa) nomenclature', *Journal of Protozoology*, *29*, 601-3
—*et al.* (1980) 'A newly revised classification of the Protozoa', *Journal of Protozoology*, *27*, 37-58
Lindberg, R.E. and Bovee, E.C. (1976) '*Chaos carolinensis*, induction of phagocytosis and cannibalism', *Journal of Protozoology*, *23*, 333-6
Lindmark, D.G. (1980) 'Energy metabolism of the anaerobic protozoan *Giardia lamblia*', *Molecular and Biochemical Parasitology*, *1*, 1-12
Ling, K-Y. and Kung, C. (1980) 'Ba^{2+} influx measures the duration of membrane excitation in *Paramecium*', *Journal of Experimental Biology*, *84*, 73-87
Listebarger, J.K. and Mitchell, L.G. (1980) 'Scanning electron microscopy of spores of the Myxosporidans *Chloromyxum trijugum* and *Chloromyxum catostomi*', *Journal of Protozoology*, *27*, 155-9
Littlepage, J. (1968) 'Plankton investigations in McMurdo Sound', *Antarctic Journal of the USA*, *3*, 162-4
Litvin, F.F., Sineshchekov, O.A. and Sineshchekov, V.A. (1978) 'Photoreceptor electric potential in the phototaxis of the alga *Haematococcus pluvialis*', *Nature*, *271*, 476-8
Lloyd, D. and Cantor, M.H. (1979) 'Subcellular structure and function in the acetate flagellates' in M. Levandowsky and S.H. Hutner (eds.), *Biochemistry and Physiology of Protozoa*, vol. 2 (Academic Press, New York), pp. 9-65
Lloyd, L. (1945) 'Animal life in sewage purification processes', *Water Pollution Control*, *44*, 119-39
Luckinbill, L.S. (1973) 'Co-existence in laboratory populations of *Paramecium aurelia* and its predator *Didinium nasutum*', *Ecology*, *54*, 1320-7
—(1974) 'The effects of space and enrichment on a predator-prey system', *Ecology*, *55*, 1142-7
—(1979) 'Selection and the r/K continuum in experimental populations of protozoa', *American Naturalist*, *113*, 427-37
Lumsden, W.H.R. and Evans, D.A. (1976) *Biology of the Kinetoplastida*, vol. 1 (Academic Press, London)

Lumsden, W.H.R. and Evans, D.A. (1979) *Biology of the Kinetoplastida*, vol. 2 (Academic Press, London)
Lushbaugh, W.B. and Pitmann, F.E. (1979) 'Microscopic observations on the filopodia of *Entamoeba histolytica*', *Journal of Protozoology*, *26*, 186-95
Lynch, J.M. and Poole, N.J. (1979) *Microbial Ecology: a Conceptual Approach* (Blackwell, Oxford)
McKanna, J. (1973a) 'Cyclic membrane flow in the ingestive-digestive system of peritrich protozoans. 1. Vesicular fusion at the cytopharynx', *Journal of Cell Science*, *13*, 663-75
—(1973b) 'Cyclic membrane flow in the ingestive-digestive system of peritrich protozoans.2. Cup-shaped coated vesicles', *Journal of Cell Science*, *13*, 677-86
Mack, W.N., Mack, J.P. and Ackerson, A.O. (1975) 'Microbial film development in trickling filters', *Microbial Ecology*, *2*, 215-26
MacKeen, P.C. and Mitchell, R.B. (1977) 'Prey capture capacity in the suctorian protozoan *Tokophrya lemnarum*', *Transactions of American Microscopical Society*, *96*, 68-75
Maguire, B. Jr. and Belk, D. (1967) '*Paramecium multimicronucleatum* transport by land snails', *Journal of Protozoology*, *14*, 445-7
Maly, E.J. (1978) 'Stability of the interaction between *Didinium* and *Paramecium*: effects of dispersal and predator time lag', *Ecology*, *59*, 733-41
Manaia, A. de C., de Souza, M.C.M., Lustosa, E. de S. and Roitman, I. (1981) '*Leptomonas lactosovorans* n.sp., a lactose-utilizing trypanosomatid: description and nutritional requirements', *Journal of Protozoology*, *28*, 124-6
Mann, K.H., Britton, R.H., Kowalczewski, A., Lack, T.J., Mathews, C.P. and McDonald, I. (1972) 'Productivity and energy flow at all trophic levels in the River Thames, England' in Z. Kajak and A. Hillbricht-Ilkowska (eds.), *Productivity Problems of Freshwaters* (Proceedings of the IBP-UNESCO Symposium, Polish Scientific Publications, Warsaw), pp. 579-96
Manton, I. (1964a) 'Further observations on the fine structure of the haptonema of *Prymnesium parvum*', *Arch.Mikrobiologie*, *49*, 315-30
—(1964b) 'Observations with the electron microscope on the division cycle in the flagellate *Prymnesium parvum* Carter', *Journal of the Royal Microscopical Society*, *83*, 317-25
—and Parke, M. (1960) 'Further observations on small green flagellates with special reference to possible relatives of *Chromulina pusilla* Butcher', *Journal of Marine Biological Assoc., UK*, *39*, 275-98
—and — (1965) 'Observations on the fine structure of two species of *Platymonas* with special reference to flagellar scales and the mode of origin of the theca', *Journal of Marine Biological Assoc., UK*, *45*, 743-54
—, Rayns, D.G., Ettl, H. and Parke, M. (1965) 'Further observations on green flagellates with scaly flagella: the genus *Heteromastix* Korshikov', *Journal of Marine Biological Assoc., UK*, *45*, 241-55
—, Sutherland, J. and Leadbeater, B.S.C. (1976) 'Further observations on the fine structure of marine collared flagellates (Choanoflagellata) from arctic Canada and west Greenland: species of *Parvicorbicula* and *Pleurasiga*', *Canadian Journal of Botany*, *54*, 1932-55
Marbach, I. and Mayer, A.M. (1971) 'Effect of electric field on the phototactic response of *Chlamydomonas reinhardii*', *Israel Journal of Botany*, *20*, 96-100
Margulis, L. (1981) *Symbiosis in Cell Evolution* (W.H. Freeman, San Francisco)

Marr, J.J. (1980) 'Carbohydrate metabolism in *Leishmania*' in M. Levandowsky and S.H. Hutner (eds.), *Biochemistry and Physiology of Protozoa*, vol. 3 (Academic Press, New York), pp. 313-40

Martinez-Palomo, A. (1982) *The Biology of* Entamoeba histolytica (Research Studies Press, John Wiley & Sons Ltd, Chichester, England)

Mast, S.O. and Doyle, W.L. (1934) 'Ingestion of fluids by *Amoeba*', *Protoplasma*, *20*, 555-60

Mattern, C.F.T., Daniel, W.A. and Honigberg, B.M. (1969) 'Structure of *Hypotrichomonas acosta* (Moskowitz) as revealed by electron microscopy', *Journal of Protozoology 16*, 668-85

Meier, R., Reisser, W. and Wiessner, W. (1980) 'Cytological studies on the endosymbiotic unit of *Paramecium bursaria* Ehr. and *Chlorella* sp. II. The regulation of the endosymbiotic algal population as influenced by the nutritional condition of the symbiotic partners', *Archiv für Protistenkunde*, *123*, 333-41

Melkonian, M. and Robenek, H. (1979) 'The eyespot of the flagellate *Tetraselmis cordiformis* Stein (Chlorophyceae)', *Protoplasma*, *100*, 183-97

Menke, W. (1962) 'Structure and chemistry of plastids', *Annual Review of Plant Physiology*, *13*, 27-44

Mignot, J-P. (1966) 'Structure et ultrastructure de quelques Euglénomonadines', *Protistologica*, *2*, 51-117

Milder, R. and Deane, M.P. (1969) 'The cytostome of *Trypanosoma cruzi* and *T. conorhini*', *Journal of Protozoology*, *16*, 730-7

Miller, S. (1968) 'The predatory behaviour of *Dileptus anser*', *Journal of Protozoology*, *15*, 313-19

Morishita, I. (1976) 'Protozoa in sewage and waste water treatment systems', *Transactions of American Microscopical Society*, *95*, 373-7

Mueller, B.D., Desser, S.S. and Haberkorn, A. (1981) 'Ultrastructure of developing gamonts of *Eimeria contorta* Haberkorn, 1971 (Protozoa, Sporozoa) with emphasis on the host-parasite interface', *Journal of Parasitology*, *67*, 487-95

Mueller, J.A. and Mueller, W.P. (1970) '*Colpoda cucullus*: a terrestrial aquatic', *American Midland Naturalist*, *83*, 1-12

Müller, M. (1970) 'The release of hydrolases by *Tetrahymena pyriformis*', *Journal of Protozoology*, *17* (Suppl.), 13

—, Baudhuin, P. and de Duve, C. (1966) 'Lysosomes in *Tetrahymena pyriformis*', *Journal of Cell Physiology*, *68*, 165-75

—, Röhlich, P. and Törö, I. (1965) 'Studies on feeding and digestion in protozoa.7. Ingestion of polystyrene latex particles and its early effect on acid phosphatase in *Paramecium micronucleatum* and *Tetrahymena pyriformis*', *Journal of Protozoology*, *12*, 27-34

—, —, Tóth, J. and Törö, I. (1963) *Lysosomes* (CIBA Foundation Symposium) (Churchill, London), pp. 201-16

Münch, R. (1970) 'Food uptake by endocytosis in *Opalina ranarum*', *Cytobiologie*, *2*, 108-22 (English summary)

Murray, J.W. (1973) *Distribution and Ecology of Living Benthic Foraminiferids* (Heinemann, London)

Neilson, A.H., Holm-Hansen, O. and Lewin, R.A. (1972) 'An obligately autotrophic mutant of *Chlamydomonas dysosmos*: a biochemical elucidation', *Journal of General Microbiology*, *71*, 141-8

Nerad, T.A. and Daggett, P-M. (1979) 'Starch gel electrophoresis: an effective method of separation of pathogenic and non-pathogenic *Naegleria* strains', *Journal of Protozoology*, *26*, 613-15

Newton, B.A. (1976) 'Biochemical approaches to the taxonomy of kinetoplastid flagellates' in W.H.R. Lumsden and D.A. Evans (eds.), *The Biology of the Kinetoplastida*, vol. 1 (Academic Press, London), pp. 405-34

Nilsson, J.R. (1977) 'On food vacuoles in *Tetrahymena pyriformis* GL', *Journal of Protozoology*, *24*, 502-7

—(1979) 'Phagotrophy in *Tetrahymena*' in M. Levandowsky and S.H. Hutner (eds.), *Biochemistry and Physiology of Protozoa*, vol. 2 (Academic Press, New York), pp. 339-79

Nisbet, B. (1974) 'An ultrastructural study of the feeding apparatus of *Peranema trichophorum*', *Journal of Protozoology*, *21*, 39-48

Noirot-Timothée, C. (1966) 'Présence simultanée de deux types de vésicules de micropinocytose chez *Cepedea dimidiata*', *C.R. Hebd. seances Acad. Sc.*, *263*, 1230-3

Noland, L.E. and Gojdics, M. (1967) 'Ecology of free-living protozoa', in T-T. Chen (ed.), *Research in Protozoology*, vol. 2 (Pergamon Press, Oxford), pp. 241-7

Oduro, K.K., Bowman, I.B.R. and Flynn, I.W. (1980) '*Trypanosoma brucei*: preparation and some properties of the multienzyme complex catalysing part of the glycolytic pathway', *Experimental Parasitology*, *50*, 240-50

Open University (1978) *Oceanography, Unit 10, the Benthic System* (Open University Press, Milton Keynes)

Orpin, C.G. and Letcher, A.J. (1978) 'Some facts controlling the attachment of the rumen holotrich protozoa *Isotricha intestinalis* and *I. prostoma* to plant particles *in vitro*', *Journal of General Microbiology*, *106*, 33-40

Oxford, A.E. (1955) 'The rumen ciliate protozoa: their chemical composition, metabolism, requirements for maintenance and culture, and physiological significance for the host', *Experimental Parasitology* (Parasitological Review, no. 6), *4*, 569-605

Page, F.C. (1967) 'Re-definition of the genus *Acanthamoeba* with descriptions of three species', *Journal of Protozoology*, *14*, 709-24

—(1976) *An Illustrated Key to Freshwater and Soil Amoebae* (Freshwater Biological Association, Windermere, England)

Parker, J.G. (1976) 'Cultural characteristics of the marine ciliate protozoan, *Uronema marinum*', *Journal of Experimental Marine Biology and Ecology*, *24*, 213-26

Patrick, R. (1949) 'A proposed biological measure of stream conditions, based on a survey of the Conestoga Basin, Lancaster County, PA', *Proc. Acad. Nat. Sci. Phila.*, *101*, 277-341

—(1950) 'Biological measure of stream conditions', *Sewage and Industrial Wastes*, *22*, 926-38

—(1961) 'A study of the number and kinds of species found in rivers in the Eastern United States', *Proc. Acad. Nat. Sci. Phila.*, *113*, 215-58

Patterson, D.J. (1979) 'On the organization and classification of the protozoan *Actinophrys sol* Ehrenberg, 1830', *Microbios*, *26*, 165-208

—(1983) Personal communication (Bristol University)

—and Hausmann, K. (1981) 'Feeding by *Actinophrys sol* (Protista, Heliozoa): 1. Light microscopy', *Microbios*, *31*, 39-55

Paulin, J.J. and Corliss, J.O. (1969) 'Ultrastructure and other observations which suggest suctorian affinities for the taxonomically enigmatic ciliate *Cyathodinium*', *Journal of Protozoology*, *16*, 216-23

Peck, R. and Hausmann, K. (1980) 'Primary lysosomes of the ciliate *Pseudomicrothorax dubius*: cytochemical identification and role in phagocytosis', *Journal of Protozoology*, *27*, 401-9

—, Pelvat, B., Bolivar, I. and Haller, G. (1975) 'Light and electron microscope observations on the heterotrich ciliate *Climacostomum virens*', *Journal of Protozoology*, *22*, 368-85

Perkins, F.O. (1976) 'Zoospores of the oyster pathogen *Dermocystidium marinum*. I. Fine structure of the conoid and other sporozoan-like organelles', *Journal of Parasitology*, *62*, 959-74

Pianka, E.R. (1970) 'On r- and K-selection', *American Naturalist*, *100*, 463-5

Picken, L.E.R. (1937) 'The structure of some protozoan communities', *Journal of Ecology*, *25*, 368-84

Pitelka, D.R. (1969) 'Fibrillar systems in protozoa' in T-T. Chen (ed.), *Research in Protozoology* vol. 3 (Pergamon Press, Oxford), pp. 333-42

Pittilo, R.M., Ball, S.J. and Hutchinson, W.M. (1980) 'The ultrastructural development of the macrogamete of *Eimeria stiedai*', *Protoplasma*, *104*, 33-41

Preer, J.R. Jr. (1975) 'The hereditary symbionts of *Paramecium aurelia*' in D.H. Jennings and D.L. Lee (eds.), *Symbiosis*, Society of Experimental Biology, Symp. 29 (Cambridge University Press), pp. 125-44

Preston, T.M. (1969) 'The form and function of the cytostome-cytopharynx of the culture forms of the elasmobranch haemoflagellate *Trypanosoma raiae* Laveran and Mesnil', *Journal of Protozoology*, *16*, 320-33

—, Davies, D.H. and King, C.A. (1982) 'Surface binding and subsequent capping of certain bacteria by *Acanthamoeba*' *Journal of Protozoology*, *29*, 318

Pringsheim, E.G. (1937) 'Assimilation of different organic substances by saprophytic flagellates', *Nature*, *139*, 196

Prusch, R.D. (1980) 'Endocytotic sucrose uptake in *Amoeba proteus* induced with the calcium ionophore', *Science*, *209*, 691-2

—and Hannafin, J.A. (1979) 'Sucrose uptake by pinocytosis in *Amoeba proteus* and the influence of external calcium', *Journal of General Physiology*, *74*, 523-35

Puytorac, P. de *et al.* (1974) 'Proposition d'une classification de Phylum Ciliophora Doflein 1901', *Comp. Rend. Acad. Sci.*, *278*, 2799-802

Rapport, D.J., Berger, J. and Reid, D.B.W. (1972) 'Determination of food preference of *Stentor coeruleus*', *Biological Bulletin*, *142*, 103-9

—and Turner, J.E. (1970) 'Determination of predator food preferences', *Journal of Theoretical Biology*, *26*, 365-72

Raven, J.A. (1970) 'The role of cyclic and pseudocyclic photophosphorylation, in photosynthetic $^{14}CO_2$ fixation in *Hydrodictyon africanum*', *Journal of Experimental Botany*, *21*, 1-16

—(1971) 'Cyclic and noncyclic photophosphorylation as energy sources for active K influx in *Hydrodictyon africanum*', *Journal of Experimental Botany*, *22*, 420-33

Rhoads, D.E. and Kaneshiro, E.S. (1979) 'Characterizations of phospholipids from *Paramecium tetraurelia* cells and cilia', *Journal of Protozoology*, *26*, 329-38

Roitman, I., Heyworth, P.G. and Gutteridge, W.E. (1978) 'Lipid synthesis by *Trichomonas vaginalis*', *Ann. Trop. Med. Parasitol.*, *72*, 583-5

Roth, L.E. (1960) 'Electron microscopy of pinocytosis and food vacuoles in *Pelomyxa*', *Journal of Protozoology*, 7, 176-85

Rudzinska, M.A. (1965) 'The fine structure and the function of the tentacle of *Tokophrya infusionum*', *Journal of Cell Biology*, *25*, 459-77

—(1970) 'The mechanism of food intake in *Topkophrya infusionum* and ultrastructural changes in food vacuoles during digestion', *Journal of Protozoology*, *17*, 626-41

—(1972) 'Ultrastructural localization of acid phosphatase in feeding *Tokophrya infusionum*', *Journal of Protozoology*, *19*, 618-29

—(1973) 'Autophagy in *Tokophrya infusionum*' in P. de Puytorac and J. Grain (eds.), *Progress in Protozoology* (4th International Congress, Université de Clermont-Ferrand), p. 354

—(1974) 'Ultrastructural localization of acid phosphatase in starved *Tokophrya infusionum*', *Journal of Protozoology*, *21*, 721-8

—(1976) 'Ultrastructure of intraerythrocytic *Babesia microti*, emphasising feeding mechanisms', *Journal of Protozoology*, *23*, 224-33

—(1980) 'Internalization of macromolecules from the medium in Suctoria', *Journal of Cell Biology*, *84*, 172-83

Ruff, M.D. and Read, C.P. (1974) 'Specificity of carbohydrate transport in *Trypanosoma equiperdum*', *Parasitology*, *68*, 103-15

Ryley, J.F. (1956) 'Studies on the metabolism of the protozoa.7. Comparative carbohydrate metabolism of eleven species of trypanosome', *Biochemical Journal*, *62*, 215-22

—(1962) 'Studies on the metabolism of the protozoa. 9. Comparative metabolism of bloodstream and culture forms of *Trypanosoma rhodesiense*', *Biochemical Journal*, *85*, 211-23

—, Bentley, M., Manners, D.J. and Stark, J.R. (1969) 'Amylopectin, the storage polysaccharide of the coccidia, *Eimeria brunetti* and *E. tenella*', *Journal of Parasitology*, *55*, 839-45

Sagar, R. and Palade, G.E. (1957) 'Structure and development of the chloroplast in *Chlamydomonas*. I.The normal green cell', *Journal of Biophysics and Biochemistry*, *3*, 463-88

Salt, G.W. (1967) 'Predation in an experimental protozoan population (*Woodruffia-Paramecium*)', *Ecological Monographs*, *37*, 113-44

—(1968) 'The feeding of *Amoeba proteus* on *Paramecium aurelia*', *Journal of Protozoology*, *15*, 275-80

—(1974) 'Predator and prey densities as controls of the rate of capture by the predator *Didinium nasutum*', *Ecology*, *55*, 434-9

—(1979) 'Density, starvation and swimming rate in *Didinium* populations', *American Naturalist*, *113*, 135-43

Sandon, H. (1924) 'Some protozoa from the soils and mosses of Spitsbergen', *Linnean Society Journal of Zoology*, *35*, 449-75

—(1927) *The Composition and Distribution of the Protozoan Fauna of the Soil* (Oliver and Boyd, Edinburgh)

—(1932) *The Food of Protozoa* (Cairo University Press)

—and Cutler, D.W. (1924) 'Some protozoa from the soils collected by the 'Quest' Expedition (1921-22)', *Linnean Society Journal of Zoology*, *36*, 1-12

Sattler, C.A. and Staehelin, L.A. (1979) 'Oral cavity of *Tetrahymena pyriformis*: a freeze fracture and high voltage electron microscopy study of the oral ribs, cytostome and forming food vacuole', *Journal of Ultrastructure Research*, *66*, 132-50

Sawyer, T.K. (1971) 'Isolation and identification of the free-living marine amoebae from Upper Chesapeake Bay, Maryland', *Transactions of American Microscopical Society*, *90*, 43-51

—(1980) 'Marine amoebae from clean and stressed bottom sediments of the Atlantic Ocean and Gulf of Mexico', *Journal of Protozoology*, *27*, 13-32

Schofield, T. (1971) 'Some biological aspects of the activated sludge plant at Leicester', *Water Pollution Control*, *70*, 32-47

Schraw, W.P. and Vaughan, G.L. (1979) '*Trypanosoma lewisi*: alterations in membrane function in the rat', *Experimental Parasitology*, *48*, 15-26

Schuster, F.L. (1979) 'Small Amebas and Ameba flagellates' in M. Levandowsky and S.H. Hutner (eds.), *Biochemistry and Physiology of Protozoa*, vol. 1 (Academic Press, New York)

Seravin, L.N. and Orlovskaja, E.E. (1973) 'Factors responsible for food selection in protozoa' in P. de Puytorac and J. Grain (eds.), *Progress in Protozoology* (4th International Congress, Université de Clermont-Ferrand), p. 371

—and — (1977) 'Feeding behaviour of unicellular animals. 1. The main role of chemoreception in the food choice of carnivorous protozoa', *Acta Protozoologica*, *16*, 309-32

Serrano, R. and Reeves, R.E. (1975) 'Physiological significance of glucose transport in *Entamoeba histolytica*', *Experimental Parasitology*, *37*, 411-16

Sharabi, Y. and Gilboa-Garber, N. (1980) 'Interactions of *Pseudomonas aeruginosa* hemaglutinins with *Euglena gracilis*, *Chlamydomonas reinhardi* and *Tetrahymena pyriformis*', *Journal of Protozoology*, *27*, 80-3

Sherman, G.B., Buhse, H.E. Jr. and Smith, H.E. (1982) 'Physiological studies on the cytophargyngeal pouch, a prey receptacle in the carnivorous macrostomal form of *Tetrahymena vorax*', *Journal of Protozoology*, *29*, 360-5

Sherman, I.W. and Hull, R.W. (1960) 'The pigment (hemozoin) and proteins of the avian malarial parasite, *Plasmodium lophurae*', *Journal of Protozoology*, 7, 409-16

Shorb, M.S. (1964) 'The physiology of Trichomonads' in S.H. Hutner (ed.), *Biochemistry and Physiology of Protozoa*, vol. 3 (Academic Press, New York), pp. 383-457

—and Lund, P.G. (1959) 'Requirement of trichomonads for unidentified growth factors, saturated and unsaturated fatty acids', *Journal of Protozoology*, *6*, 122-30

Sieburth, J.M. (1979) *Sea Microbes* (Oxford University Press, New York)

Sleigh, M.A. (1964) 'Flagellar movement of the sessile flagellates *Actinomonas*, *Codonosiga*, *Monas* and *Poteriodendron*', *Quarterly Journal of Microscopical Science*, *105*, 405-14

—(1973) *The Biology of Protozoa* (Edward Arnold, London)

—and Barlow, D. (1976) 'Collection of food by *Vorticella*', *Transactions of American Microscopical Society*, *95*, 482-6

Small, E.B. (1973) 'A study of ciliate protozoa from a small polluted stream in East-central Illinois', *American Zoologist*, *13*, 225-30

Smillie, R.M. (1968) 'Enzymology of *Euglena*' in D.E. Buetow (ed.), *The Biology of Euglena*, vol. 2 (Academic Press, New York), pp. 43-50

Smith, H.G. (1973a) 'The Signy Island terrestrial reference sites.III. Population ecology of *Corythion dubium* in Site 1', *British Antarctic Survey Bulletin, No. 33-4*, 123-235

—(1973b) 'The ecology of protozoa in Chinstrap Penguin guano', *British Antarctic Survey Bulletin, no. 35*, 33-50

—(1973c) 'The temperature relations and bi-polar geography of the ciliate genus *Colpoda*', *British Antarctic Survey Bulletin, no. 37*, 7-13

—(1978) 'The distribution and ecology of terrestrial protozoa of sub-antarctic and maritime antarctic islands', *British Antarctic Survey Scientific Reports, no. 95*

Smith, J.E. and Sinden, R.E. (1980) 'A technique for the culture of *Nosema algerae* in primary cultures of rat brain', *Journal of Protozoology*, *27*, 59A

Smith, T.P. (1978) 'Distribution of benthic ciliophora in Point Mugu Lagoon, Southern California', *American Zoologist*, *81*, 660

Soldo, A.T. and van Wagtendonk, W.J. (1967) 'An analysis of the nutritional requirements for fatty acids in *Paramecium aurelia*', *Journal of Protozoology*, *14*, 596-600

Sprague, V. and Vernick, S.H. (1974) 'Fine structure of the cyst and some sporulation stages of *Ichthyosporidium* (Microsporida)', *Journal of Protozoology*, *21*, 667-77

Stewart, W.D.P. (1974) *Algal Physiology and Biochemistry* (Blackwell, Oxford) chs. 4-6

Stockem, W. (1973) 'Morphological and cytochemical aspects of endocytosis and intracellular digestion in amoebae' in P. de Puytorac and J. Grain (eds.), *Progress in Protozoology*, (4th International Congress, Université de Clermont-Ferrand), p. 402

Stout, J.D. (1955) 'Environmental factors affecting the life history of three species of *Colpoda* (Ciliata)', *Transactions of Royal Society, New Zealand*, *82*, 1165-88

—(1962) 'An estimation of microfaunal populations in soils and forest litter', *Journal of Soil Science*, *13*, 314-20

—(1970) 'The bacteria and protozoa of some soil samples from Scoresby Land', *Meddelelseer om Grønland*, *184*, no. 11

Sugden, B. (1950) 'A study of the feeding and excretion of the ciliate *Carchesium*, in relation to the clarification of sewage effluent', PhD *thesis, University of Leeds*

—(1953) 'The cultivation and metabolism of oligotrich protozoa from the sheep's rumen', *Journal of General Microbiology*, *9*, 44-53

—and Lloyd, L. (1950) 'The clearing of turbid waters by means of the ciliate *Carchesium*', *Water Pollution Control*, *49*, 16-23

—and Oxford, A.E. (1952) 'Some cultural studies with holotrich ciliate protozoa of the sheep's rumen', *Journal of General Microbiology*, *7*, 145-53

Swedmark, B. (1964) 'The interstitial fauna of marine sand', *Biological Reviews*, *39*, 1-42

Tai, L-S. and Skogsberg, T. (1934) 'Studies on the Dinophysidae, marine armored dinoflagellates of Monterey Bay, California', *Archiv für Protistenkunde*, *82*, 380-482

Tamm, S.L. (1982) 'Flagellated ectosymbiotic bacteria propel a eukaryote cell',

Journal of Cell Biology, *94*, 697-709

Tanner, W., Loffler, W. and Kandler, O. (1969) 'Cyclic photophosphorylation *in vivo* and its relation to photosynthetic CO_2 fixation', *Plant Physiology, Lancaster*, *44*, 422-8

Taylor, D.L. and Seliger, H.H. (1979) 'Toxic dinoflagellate blooms' in *Developments in Marine Biology*, vol. 1 (Elsevier/North Holland, New York)

Taylor, F.J.R. (1974) 'Implications and extensions of the serial endosymbiosis theory of the origin of eukaryotes', *Taxon*, *23*, 229-258

Taylor, M.B. and Gutteridge, W.E. (1980) 'Apparent presence of glycosomes in *Trypanosoma cruzi*', *Parasitology*, *81*, Pt 2, iv

Taylor, W.D. (1978) 'Growth responses of ciliate protozoa to the abundance of their bacterial prey', *Microbial Ecology*, *4*, 207-14

Thomas, G.J. (1960) 'Metabolism of the soluble carbohydrates of grasses in the rumen of sheep', *Journal of Agricultural Science*, *54*, 360-72

Tietjen, J.H. (1971) 'Ecology and distribution of deep-sea meiobenthos off North Carolina', *Deep-Sea Research*, *18*, 941-57

Trueman, E.R. (1975) *The Locomotion of Soft-bodied Animals* (Edward Arnold, London), pp. 122-4

Tucker, J.B. (1968) 'Fine structure and function of the cytopharyngeal basket in the ciliate *Nassula*', *Journal of Cell Science*, *3*, 493-514

—(1972) 'Microtubule arms and propulsion of food particles inside a large feeding organelle in the ciliate *Phascolodon vorticella*', *Journal of Cell Science*, *10*, 883-903

—(1974) 'Microtubule arms and cytoplasmic streaming and microtubule bending and stretching of intertubule links in the feeding tentacle of the suctorian ciliate *Tokophrya*', *Journal of Cell Biology*, *62*, 424-37

Tuffrau, M., Tuffrau, H. and Genermont, J. (1976) 'La réorganisation infraciliaire au cours de la conjugaison et l'origine du primordium buccal dans le genre *Euplotes*', *Journal of Protozoology*, *23*, 517-23

Uyemura, M. (1936) 'Biological studies of thermal waters in Japan, 4', *Ecological Studies*, *2*, 171

Verni, F. and Rosati, G. (1980) 'Preliminary survey of the morphology and cytochemistry of polysaccharide reserves in ciliates'. *Protistologica*, *16*, 427-34

Vickerman, K. (1965) 'Polymorphism and mitochondrial activity in sleeping sickness trypanosomes', *Nature*, *208*, 762-6

—(1969a) 'The fine structure of *Trypanosoma congolense* in its bloodstream phase', *Journal of Protozoology*, *16*, 54-69

—(1969b) 'On the surface coat and flagellar adhesion in trypanosomes', *Journal of Cell Science*, *5*, 163-93

—(1976) 'The diversity of kinetoplastid flagellates' in W.H.R. Lumsden and D.A. Evans (eds.), *Biology of the Kinetoplastida*, vol. 1 (Academic Press, London), pp. 5-8

—(1978) 'Antigenic variation in trypanosomes', *Nature*, *273*, 613-17

—(1982) 'Parasitic protozoa: aspects of the host-parasite interface' in D.F. Mettrick and S.S. Deoser (eds.) *Parasites – their World and Ours* (Elsevier Biomedical Press, New York), pp. 43-52

—, Barry, J.D., Hajduk, S.L. and Tetley, L. (1980) in H. van den Bossche (ed.), *The Host-Invader Interplay* (Elsevier/North Holland, Amsterdam), pp. 179-90

—and Preston, T.M. (1976) 'Comparative cell biology of the kinetoplastid

flagellates' in W.H.R. Lumsden and D.A. Evans (eds.), *Biology of the Kinetoplastida*. vol. 1 (Academic Press, London), pp. 60-3
Vickerman, K. , and Tetley, L. (1977) 'Recent ultrastructural studies on trypanosomes', *Ann.Soc.belg. Méd. trop*., *57*, 441-55
Villarreal, E., Canale, R.R. and Akcasu, Z. (1977) 'Transport equations for a microbial predator-prey community', *Microbial Ecology*, *3*, 131-42
Voorheis, H.P. (1980) 'Fatty acid uptake by bloodstream forms of *Trypanosoma brucei* and other species of Kinetoplastida', *Molecular and Biochemical Parasitology*, *1*, 177-86
Wallis, O.C. and Coleman, G.S. (1967) 'Incorporation of ^{14}C-labelled components of *Escherichia coli* and of amino acids by *Isotricha intestinalis* and *Isotricha prostoma*', *Journal of General Microbiology*, *49*, 315-23
Wang, C.C. (1928) 'Ecological studies of the seasonal distribution of protozoa in a freshwater pond', *Journal of Morphology and Physiology*, *46*, 431-78
—, Weppelman, R.M. and Lopez-Ranos, B. (1975) 'Isolation of amylopectin granules and identification of amylopectin phosphorylase in the oocysts of *Eimeria tenella*', *Journal of Protozoology*, *22*, 560-4
Watson, J.M. (1945) 'Mechanism of bacterial flocculation by protozoa', *Nature*, *155*, 271
Webb, M.G. (1956) 'An ecological study of brackish water ciliates'. *Journal of Animal Ecology*, *25*, 148-75
—(1961) 'The effects of thermal stratification on the distribution of benthic protozoa in Esthwaite Water', *Journal of Animal Ecology*, *30*, 137-51
Weidner, E. (1976) 'Ultrastructure of the peripheral zone of a *Glugea*-induced xenoma', *Journal of Protozoology*, *23*, 234-8
Weinbach, E.C., Harlow, D.R., Takeuchi, T., Diamond, L.S., Claggett, C.E. and Kon, H. (1976) 'Aerobic metabolism of *Entamoeba histolytica:* facts and fallacies' in B. Sepulveda and L.S. Diamond (eds.), *Proceedings of the International Conference of Amoebiasis* (Instituto Mexicano de Seguro Social, Mexico City), pp. 190-203
Wessenberg, H.S. (1961) 'Studies on the life cycle and morphogenesis of *Opalina*', *University of California Publications in Zoology*, *61*, 315-70
—(1978) 'Opalinata' in J.P. Kreier (ed.), *Parasitic Protozoa*, vol. 2 (Academic Press, New York), pp. 551-81
—and Antipa, G. (1970) 'Capture and ingestion of *paramecium* by *Didinium nasutum*', *Journal of Protozoology*, *17*, 250-70
Whitfield, P.J. (1979) *The Biology of Parasitism* (Edward Arnold, London), ch. 1
Williams, A.G. (1979) 'Exocellular carbohydrase formation by rumen holotrich ciliates', *Journal of Protozoology*, *26*, 665-72
Williams, N.E. and Bakowska, J. (1982) 'Scanning electron microscopy of cytoskeletal elements in the oral apparatus of *Tetrahymena*', *Journal of Protozoology*, *29*, 382-9
Wilson, P.A.G. and Fairbairn, D. (1961) 'Biochemistry of sporulation in oocysts of *Eimeria acervulina*', *Journal of Protozoology*, *8*, 410-16
Wilson, R.J. (1982) 'How the malarial parasite enters the red blood cell', *Nature*, *295*, 368-9
Winterbourn, M.J. and Brown, T.J. (1967) 'Observations on the faunas of two warm streams in the Taupo thermal region', *New Zealand Journal of Marine and Freshwater Research*, *1*, 38-50

Wise, D.L. (1959) 'Carbon nutrition and metabolism in *Polytomella caeca*', *Journal of Protozoology*, *6*, 19-23

Witkamp, M. (1966) 'Decomposition of leaf litter in relation to environment, microflora and microbial respiration', *Ecology*, *47*, 194-201

Wood, H.G. (1977) 'Some reactions in which inorganic pyrophosphate replaces ATP and serves as an energy source', *Federation Proceedings*, *36*, 2197-205

Wu, C. and Hogg, J.F. (1956) 'Free and non-protein amino acids of *Tetrahymena pyriformis*', *Archives of Biochemistry and Biophysics*, *62*, 70-7

Yamin, M.A. (1978) 'Axenic culture of the cellulolytic flagellate *Trichomitopsis termopsidis* (Cleveland) from the termite', *Journal of Protozoology*, *25*, 535-8

Yocom, H.B. (1934) 'Observations on the experimental adaptation of certain freshwater ciliates to sea water'. *Biological Bulletin*, *67*, 273-6

Yonge, C.M. (1928) 'Feeding mechanisms in the invertebrates', *Biological Reviews*, *3*, 21-76

—(1954) 'Feeding mechanisms in the Invertebrata', *Tabulae Biologicae*, *21*, Part 3, no. 22

INDEX

Manufactured by Amazon.ca
Acheson, AB